MW00570813

Property Inspector's Guide to Codes, Forms, and Complaints

Property Inspector's Guide to Codes, Forms, and Complaints

Linda Pieczynski

THOMSON

DELMAR LEARNING

Australia · Canada · Mexico · Singapore · Spain · United Kingdom · United States

Property Inspector's Guide to Codes, Forms, and Complaints
Linda Pieczynski

Vice President, Technology Professional Business Unit:
Gregory L. Clayton

Product Development Manager:
Ed Francis

Development:
Sarah Boone

Director of Marketing:
Beth A. Lutz

Marketing Specialist:
Marissa Maiella

Art Director:
Robert Plant

Marketing Coordinator:
Jennifer Stall

Director of Production:
Patty Stephan

Production Manager:
Andrew Crouth

Content Project Manager:
Kara DiCaterino

Director of Technology:
Tom Smith

Technology Specialist:
Jim Ormsbee

COPYRIGHT © 2007 Thomson Delmar Learning, a division of Thomson Learning Inc. All rights reserved. The Thomson Learning Inc. logo is a registered trademark used herein under license.

Printed in the United States of America
1 2 3 4 5 XX 08 07 06

For more information contact Thomson Delmar Learning Executive Woods
5 Maxwell Drive, PO Box 8007, Clifton Park, NY 12065-8007

Or find us on the World Wide Web at www.delmarlearning.com

ALL RIGHTS RESERVED. No part of this work covered by the copyright hereon may be reproduced in any form or by any means—graphic, electronic, or mechanical, including photocopying, recording, taping, Web distribution, or information storage and retrieval systems—without the written permission of the publisher.

For permission to use material from the text or product, contact us by
Tel.   (800) 730-2214
Fax   (800) 730-2215
www.thomsonrights.com

Library of Congress Cataloging-in-Publication Data

Pieczynski, Linda Sucher.
  Property inspector's guide to codes, forms, and complaints/Linda Pieczynski.
    p. cm.
  Includes index.
  ISBN 1-4180-1609-8
  1. Building laws—United States.
2. Building inspectors—United States—Handbooks, manuals, etc.
I. Title.

KF5701.P54 2006
343.73'07869—dc22

2006022619

NOTICE TO THE READER

Publisher does not warrant or guarantee any of the products described herein or perform any independent analysis in connection with any of the product information contained herein. Publisher does not assume, and expressly disclaims, any obligation to obtain and include information other than that provided to it by the manufacturer.

The reader is expressly warned to consider and adopt all safety precautions that might be indicated by the activities herein and to avoid all potential hazards. By following the instructions contained herein, the reader willingly assumes all risks in connection with such instructions.

The publisher makes no representation or warranties of any kind, including but not limited to, the warranties of fitness for particular purpose or merchantability, nor are any such representations implied with respect to the material set forth herein, and the publisher takes no responsibility with respect to such material. The publisher shall not be liable for any special, consequential, or exemplary damages resulting, in whole or part, from the readers' use of, or reliance upon, this material.

Contents

The inspiration for the book came about because of my previous experience as an assistant state's attorney. As a prosecutor, I always had a sample complaint book to use whenever I had to draft a criminal indictment. Police officers used the same form book to write criminal complaints. When I began prosecuting municipal ordinances as a village attorney, I realized that property maintenance inspectors had to write the same types of complaints, but without the kind of guidance that police officers and prosecutors took for granted. As I traveled around the country training inspectors on the legal aspects of code administration for the International Code Council, I became convinced that a book with sample forms, complaints, checklists, and other helpful guides for inspectors could make their job easier and more efficient, and enhance enforcement of the property maintenance code. Municipal attorneys could also benefit from such a book, as they are used to relying on publications for forms which can be adapted to their needs. So, I have written this guide with the hope that it will aid inspectors and prosecutors as they perform their duties to enhance the quality of life in their communities.

All of the scenarios in the forms in this book are fictional along with the addresses. I have used the names of co-workers to make the scenarios realistic, and to honor their contribution to my knowledge about the field of code inspection. They have graciously given me their permission to include them in this book.

Linda S. Pieczynski is engaged in the private practice of law at the firm she founded in 1984 in Hinsdale, Illinois. Her practice has a special emphasis on municipal prosecution in the areas of the property maintenance, building, zoning, and fire prevention codes and quasi-criminal ordinance violations. She represents a number of municipalities in DuPage County, Illinois. Prior to entering the private practice of law, Ms. Pieczynski was an Assistant State's Attorney in DuPage County, Illinois where she eventually became Deputy Chief of the Criminal Division.

Ms. Pieczynski is an instructor for the International Code Council and has prepared building code officials for certification and taught the Legal Aspects of Code Administration course in Arizona, Colorado, Florida, Illinois, Indiana, Kansas, Kentucky, Michigan, Minnesota, Missouri, Ohio, Oklahoma, Massachusetts, New Hampshire, New Jersey, Maryland, Oregon, Pennsylvania, Tennessee, Virginia, West Virginia, and for the cities of Washington, D.C. and Phoenix. Presentations have also included the BOCA International's annual and mid-year conventions, in Norfolk, Virginia, Overland Park, Kansas and Lexington, Kentucky. She has made presentations at BOCA and ICC Housing Symposiums. Ms. Pieczynski served as the Technical Advisor and Technical Reviewer for the on-line course entitled "Right of Entry" from the 2000 International Building Code.

In addition to her ICC seminars, Ms. Pieczynski has given presentations on the "Building Blocks of Code Enforcement", "Legal Liability of Code Officials", "Search and Seizure Issues for Code Officials", "Multi-Family Housing Issues", "Property Maintenance Issues", "43 Mistakes to Avoid in Code Enforcement" and "Effective Enforcement of the Property Maintenance Code".

Among Ms. Pieczynski's previous publications are Illinois Criminal Practice and Procedure 2d, by Thomson/West, 2005, and Roll Call News, a bimonthly newsletter for law enforcement agencies and criminal attorneys. She has served as the editor of the second and third editions of the Legal Aspects of Code Administration published by the International Code Council.

I would like to thank the people who have lent their names as fictitious complainants and defendants in this work including inspectors Trevor Bishop, John Black, Karyn Byrne, Stacey Crockatt, John Fincham, Tim Halik, Don Lay, Rob McGinnis, Bob Meyer, Joan Rogers, Chuck Schmidt, and Keith Steiskal, my friend, Sandi Mueller, who unwittingly started me on this path, Chief Steve Herron of the Woodridge Police Department, Chief Pat Kenny of the Hinsdale Fire Department and Deputy Chief Pat Foley of the Willowbrook Police Department. Trevor, Karyn, Don, and fire inspector Tim Kelly answered my questions about technical issues regarding property maintenance inspections which really helped me to formulate various complaints. I want to thank my family members who agreed to let me use their names for this enterprise, often as defendants, including my parents, Larry and Joan Sucher, my brothers, Mark and Keith, my sister, Lauren, her husband, Joe, and my nephews, Tim and Kyle.

A special acknowledgment must be given to Mayor William Murphy, the mayor of Woodridge, Illinois, who encouraged me to use Woodridge as the setting for these fictional violations. I want to let him know how much I appreciate that suggestion. His support for code enforcement has done much to make Woodridge the fine community that it is. I also want to thank the code department of the Village of Woodridge for allowing me to use its inspection form for multi-family housing and the building department of the Village of Hinsdale for allowing me adapt its condemnation placards for this book. I would like to thank my colleagues at the International Code Council, especially Kathleen Mihelich, who have given me the opportunity to learn so much about the enforcement of codes on a national basis during my teaching assignments.

A special thanks goes to Karyn Byrne, my dear friend, and the best inspector I know. She has given me so much encouragement on this project, served as my sounding board, made suggestions, reviewed countless forms, and taught me more about code enforcement than anybody I know.

Lastly, I need to thank my husband, Alan, for his patience while I labored over this project and spread paperwork all over the house.

Thomson Delmar Learning and the author would also like to thank the following reviewers for their valuable suggestions and expertise:

Don Lay, Village of Hinsdale, Code Enforcement

Jeff Burton, Institute for Business and Home Safety

Karyn Byrne, City of Warrenville, Code Enforcement Representatives, Inc.

Phil Rhoads, City of Freer, Director of Building and Zoning

Phil Seyboldt, City of Bedford, Building Commissioner

Tim Halik, Village of Willowbrook

William Penn, Metro Davidson Co.

*T*he forms in this book are designed to be used in conjunction with the International Property Maintenance Code, the most popular model code in the United States for enforcing property maintenance violations. Although the book is based on the 2006 and 2003 versions of the Code, the forms are easily adaptable to any version used by the code official. The forms can also be used with any municipal code that addresses property maintenance issues, because they deal with the most common types of violations that face the property maintenance inspector. The book also contains sample forms for communities that rely solely on their nuisance ordinances to enforce property maintenance offenses.

The International Property Maintenance Code (IPMC) prescribes minimum maintenance standards for all structures and premises for basic equipment and facilities for light, ventilation, occupancy limits, heating, plumbing, electricity, for safety from fire, for space, use, and location, and for safe and sanitary maintenance for all structures and premises now in existence. The property maintenance inspector's job is to enforce these minimum requirements. This often requires quite a bit of paper work on the part of the inspector. The time required to prepare this paper work is often a deterrent to effective enforcement because the inspector might be reluctant to take time away from field inspections to draft documents. The purpose of this book is to help the inspector draft the necessary paperwork for enforcement in a timely and accurate manner, thereby allowing the inspector to not only pursue enforcement but have more time to do inspections.

STRUCTURE OF THE BOOK

The structure of this book closely follows the format of the IPMC and uses the section numbers from the 2006 and 2003 versions of the Code in numerical order. For example, to find a form which will help the inspector prepare a complaint for Accumulation of Rubbish and Garbage, IPMC 307.1, the inspector would look in Chapter Three, Rubbish and Garbage Violations, Sample Complaint 307.1, just like in the IPMC.

This book contains several different types of forms for adaptation by the inspector. Chapter One contains forms every inspector needs to enforce the Code, e.g., for conducting inspections, preparing notices of violations or administrative search warrants, writing complaints for disobeying a code official, posting unsafe structures, and condemning property. Chapters Three through Seven contain sample complaints for every violation in the International Property Maintenance Code so the inspector does not have to spend time creating his or her own and to keep from making mistakes. Chapter Eight contains sample complaints which can be used by inspectors to charge nuisance violations. It is the only chapter not based on the IPMC.

In addition to sample forms and complaints, the book contains checklists and tables for the inspector to use. The purpose of the checklists is to help the inspector make sure that all the necessary steps have been taken at a particular stage of enforcement, for example, right of entry issues in connection with the enforcement of IPMC 104.4, or in drafting a complaint, IPMC 106. The tables are to help the inspector analyze concepts such as ownership, overcrowding, requirements for light and ventilation, etc. Chapter Two contains checklists and tables to help the inspector analyze many of the critical definitions in the IPMC and to determine who is responsible for a violation.

The checklists and tables are contained in the same chapter in which the concepts exist in the IPMC. For example, the checklist for "Who is the Owner?" is contained in Chapter Two, the same chapter in which "owner" is defined in the IPMC. The table for overcrowding is in Chapter Four, where the overcrowding sample complaint exists.

Each Chapter begins with a Table of Contents of that chapter's forms, complaints, checklists, and/or tables so the inspector can quickly find what he or she needs.

USING THE FORMS

The forms and complaints in this book are designed to provide a standard format for letters, notices, complaints and other documents. The part of the form that does not change from case to case is in regular type. The words in italic type are examples of how the form may be used. Therefore, the inspector should substitute specific information for the italic type in the form. In the CD-ROM that accompanies the book, the example language is removed and the form becomes a fill-in document for the inspector.

The following illustrates how this appears in the text:

Form with sample language:

The undersigned says that on or about *July 18, 2006,* at or about *2:00 p.m.* the Defendant did unlawfully commit the offense of **Failure to Maintain Interior Structure** in violation of *2006* **IPMC-305.1** as amended and adopted by reference in Section *8-1J-1(A)* of the *Village of Woodridge* Code, in that said Defendant, the owner* of *6560 Hollywood Blvd., Woodridge, IL,* failed to keep the interior of the structure in good repair, structurally sound and in a sanitary condition in that *the rafters in the attic are water damaged and rotten, there are holes in the living room walls, and the kitchen is filled with moldy dishes and food.*

The following illustrates how this appears on the CD-ROM:

Form without sample language:

The undersigned says that on or about _____, at or about _____. the Defendant did unlawfully commit the offense of **Failure to Maintain Interior Structure** in violation of _____ **IPMC-305.1** as amended and adopted by reference in Section _____ of the _____ Code, in that said Defendant, the owner* of _____, failed to keep the interior of the structure in good repair, structurally sound and in a sanitary condition in that _____.

Sometimes there is alternative language, e.g., owner or occupant, which may be used depending on the facts of the situation or special directions for the code official. This has been designated by the use of asterisks, *, **, etc. The alternative language is located at the bottom of the form.

The undersigned says that on or about *July 18, 2006,* at or about *2:00 p.m.* the Defendant did unlawfully commit the offense of **Failure to Maintain Interior Structure** in violation of **2006 IPMC-305.1** as amended and adopted by reference in Section *8-1J-1(A)* of the *Village of Woodridge* Code, in that said Defendant, the owner* of *6560 Hollywood Blvd., Woodridge, IL,* failed to keep the interior of the structure in good repair, structurally sound and in a sanitary condition in that *the rafters in the attic are water damaged and rotten, there are holes in the living room walls, and the kitchen is filled with moldy dishes and food.*

*Or, occupant

See how in the above example the alternative language is located at the bottom of the form. "Owner*" is the first alternative. "*Occupant" appears at the end of the form and is the second alternative that can be used depending on the facts of the case.

It is highly recommended that the standard language in regular type not be changed or paraphrased. The standard language is based on the exact words used in the Code. Because words have very specific language in the law and are often defined by the Code, those are the words that should be used. The inspector can use his or her own words in place of the italicized example e.g., *"the rafters in the attic are water damaged and rotten, there are holes in the living room walls, and the kitchen is filled with moldy dishes and food"* to describe the specific violation facing the inspector.

Comments or notes have been included, where appropriate, to assist the code official in preparing these documents. For example, the sample complaint for Failure to Maintain Interior Structure contains the following comment:

COMMENT: An occupant is responsible for areas he or she occupies or controls. An owner is responsible for the shared or public areas of a structure containing a rooming house, housekeeping units, a hotel, a dormitory, two or more dwelling units, or two or more nonresidential occupancies and the exterior property.

ADAPTING THE FORMS

If the inspector uses an earlier version of the International Property Maintenance Code, the forms can be used simply by changing the section number of the violation in the complaint to the one used in the older version because most of the time the language does not change. However, the inspector should check the language of the earlier version and compare it to the 2006 and 2003 Codes so the words in the form can be adjusted if necessary.

KEY PARTS OF A COMPLAINT

The sample complaints contain the necessary elements required to charge a violation, and to provide the defendant with enough information to prepare a defense. The following diagram illustrates the elements of a sample complaint.

KEY PARTS OF A COMPLAINT

SAMPLE COMPLAINT

ACCUMULATION OF RUBBISH OR GARBAGE—2006 IPMC-307.1

State and county of violation

STATE OF *ILLINOIS*
COUNTY OF DUPAGE

Name of Plaintiff, Municipality

VILLAGE OF WOODRIDGE
v.

Name, Address of Defendant

NAME: *SANDI MUELLER*
ADDRESS: *4110 Winthrop Blvd*
CITY: *Woodridge, Illinois 60517*

Date and time of violation

The undersigned says that on or about *November 18, 2006*, at or about *3:00 p.m.*

Name of violation

the Defendant did unlawfully commit the offense of **Accumulation of Rubbish or Garbage**

Section Number of Violation and Year of IPMC Adoption

in violation of the *2006* **IPMC-307.1**

Adoption by Reference Language and Section Number

as amended and adopted by reference in Section *8-1J-1(A)* of the

Name of Municipal Code

Village of Woodridge Code,

Connection between Defendant and the offense, i.e., owner

in that said Defendant, the owner of

Location of Violation

4110 Winthrop Blvd., Woodridge, IL

(Continued)

ACCUMULATION OF RUBBISH OR GARBAGE—2006 IPMC-307.1 (*CONTINUED*)

Words constituting violation and "*", "**", and "***" signifies there is alternative language

failed to keep the interior* of the premises free from the accumulation of rubbish** being combustible waste material*** in the interior of the structure in that

Description of specifics of the violation

the master bedroom and living room were filled with newspapers, cartons, boxes and magazines.

Signature of the complainant

Stacey Crockatt
Complainant

Verification for notary public

Sworn to and Subscribed before Me
This *8th* Day of *December, 2006*

_____

Notary Public

Alternative language provisions

  *Or, exterior.

 **Or, garbage.

***Or, being animal or vegetable waste.

Helpful comments for inspector

COMMENT: Inspectors have a tendency to substitute the words junk or debris for rubbish or garbage. Because rubbish and garbage are strictly defined in the IPMC, only those words should be used. See Section VI in Chapter Two to help determine whether the substance is rubbish or garbage.

CAUTION

Whenever inspectors prepare documents which might eventually be used in court, they should consult with the attorney who represents the municipality or other legal entity in court for legal advice. Many states and local governmental bodies have adopted ICC model codes as their state statutes or local ordinances but with several amendments. Most states have specific laws governing the demolition of buildings. Court rulings can effect enforcement procedure and the format of complaints.

For example, in citing the section number of the violation, the inspector must find out what is legally required in the local jurisdiction. When a code is adopted by reference, it is part of the local jurisdiction's official code. In some jurisdictions, it is sufficient to recite the section number of the IPMC directly from the code. In other jurisdictions, a judge may require that not only the section number of the IPMC be recited in the body of the complaint but also the section number in which the IPMC is actually adopted by the jurisdiction, e.g., *2006* **IPMC-307.1** as amended and adopted by reference in Section *8-1J-1(A)* of the *Village of Woodridge* Code. Still further, a court may require an additional citation to a section if that particular provision of the IPMC has been amended by the jurisdiction, e.g., Section 307.1, as amended in Section *8-1J-1(B)* of the *Village of Woodridge* Code, of the *2006* **IPMC** as amended and adopted by reference in Section *8-1J-1(A)* of the *Village of Woodridge* Code. Other courts may require that the full name of the code be spelled out, e.g., *2006* International Property Maintenance Code. The code inspector should consider recommending that the format for how the code is to be cited be set forth in the adopting ordinance.

New rulings interpreting the law are issued every day by the courts and what was accepted procedure one day becomes unacceptable the next. Code enforcement does not have a large body of law to rely on as precedent because few cases make it to the appellate courts. Before relying on this book, the inspector must make sure that the forms are suitable under the inspector's local statutes or ordinances.

SAMPLE COMPLAINT

IPMC 302.1   FAILURE TO MAINTAIN EXTERIOR PROPERTY

SAMPLE COMPLAINT

FAILURE TO MAINTAIN EXTERIOR PROPERTY—IPMC 302.1

STATE OF *ILLINOIS*
COUNTY OF DUPAGE
VILLAGE OF WOODRIDGE   ← | Name of the legal entity prosecuting the case. |

v.

NAME: ↓*ROBERT MEYER* | Name of the responsible party |

ADDRESS: ↑*4617 Carousel St.* | Address where the defendant can be served with the violation. |

CITY: *Woodridge, Illinois 60517*

The undersigned says that on or about ↓*April 2, 2006*, at or about | Date and time inspector observed violation |

3:00 p.m. the Defendant did unlawfully commit the offense of ↓**Failure** | Name of the violation |

to Maintain Exterior Property in violation of ↑**IPMC-302.1** of the | Section number of the violation |

ordinances of the *Village of Woodridge*, in that said Defendant, the | Location of the violation |

owner of ↓*4617 Carousel St., Woodridge, Illinois*, failed to ↑maintain the | Description of the violation based on the Code language |

exterior of the property in a clean, safe and sanitary condition in

that ↑*there were piles of garbage, rusted machine parts and dead tree limbs* | Description of the nature of the violation to enable defendant to defend the charge. |

in the back yard.

$$\mathcal{Joan\ Rogers}$$

↑—————————
Complainant | Signature of the inspector. |

Sworn to and Subscribed before Me

↑This *30th* Day of *April, 2006* | Date on which inspector signs complaint. |

————————————————

↑Notary Public | If local code requires verification, signature and seal of notary public. |

Administration of the Code

CHAPTER 1 is the administrative section for the 2006 International Property Maintenance Code. It details the procedural aspects of the code. The chapter describes how the code applies to the maintenance of structures and premises. It creates the department of property main- tenance, which has authority over code enforcement, and discusses the liability of code officials. It sets forth in detail the various duties and powers of the code official, and explains the process by which the code is modified. All of the steps necessary to gain compliance are present in this chapter, including notices of violations, prosecution, penalties, and abatement remedies. The chapter deals with unsafe structures and equipment and emergency measures that the code official can take, including demolition. Finally, the appeals process is reviewed.

IPMC 103   Creation of the Department of Property Maintenance Inspection

SAMPLE FORM

APPOINTMENT OF CODE OFFICIAL—IPMC 103.2

I, *Mayor William Murphy*, the chief appointing authority of the jurisdiction, being the *Village of Woodridge*, do hereby appoint *John Black* as the code official, being the executive official in charge of the Department of Property Maintenance.

Date: *March 1, 2006*

Mayor William Murphy

Chief Appointing Authority

of the *Village of Woodridge*

COMMENT: It is important to have these types of forms in the municipality's official records so a case cannot be challenged on the basis that the person had no authority to act.

SAMPLE FORM

APPOINTMENT OF DEPUTIES—IPMC 103.3

I, *John Black*, the code official for the *Village of Woodbridge*, do hereby appoint *Karyn Byrne* as deputy code official* for the *Village of Woodbridge*.

Date: *March 2, 2006*

John Black

Code Official

Approved: *Mayor William Murphy*

Chief Appointing Authority

of the *Village of Woodridge*

*Or, technical officer, inspector, or employee

IPMC 104   Duties and Power of the Code Official Including Right of Entry

SAMPLE FORM

INSPECTION RECORD—IPMC 104.3

Location of Property: _____

PIN: _____

Owner: _____ D.O.B. _____

Address: _____ City, State: _____ Zip: _____

E-mail: _____ Office #:_____ Cell: _____

FAX: _____

Co-Owner: _____ D.O.B. _____

Address: _____ City, State: _____ Zip: _____

E-mail: _____ Office #:_____ Cell: _____

FAX: _____

Occupant(s): _____ D.O.B. _____

Address: _____ City, State: _____ Zip: _____

E-mail: _____ Office #:_____ Cell: _____

FAX: _____

Corporate or L.L.C. Owner?   Yes [ ]   No [ ]

❑ Corporation – Corporate No. _____   ❑ Limited Liability Company – LLC No. _____

   Registered Agent: _____

   Address:_____

   City, State: _____ Zip: _____

Land Trust?   Yes [ ]   No [ ]

❑ Land Trust, Trustee _____

   Person with Power of Direction: _____

   Address:_____

   City, State: _____ Zip: _____

   Beneficiaries: _____

   Address:_____

   City, State: _____ Zip: _____

Ownership verified: By person's admission _____

                       By title search _____

                       Other (describe) _____

                                                                (Continued)

INSPECTION RECORD—IPMC 104.3 (*CONTINUED*)

Nature of Property: ___ Single Family Residential

___ Multi-Family Residential

___ Commercial

___ Vacant Land

___ Industrial

___ Other_____

Complainant:_____

Inspection Date: _____ Time: _____

Summary of Findings: _____

_____

_____

_____

Name of violation(s)   Code Section(s) violated:

_____   _____

_____   _____

_____   _____

Photos: Yes [ ]   No [ ]   Video: Yes [ ]   No [ ]

Contact with owner: Yes [ ]   No [ ]

If yes, summary of conversation _____

_____

_____

_____

Notice: Date sent _____ (Attach copy of Notice)

Method of Service: _____ Personal Service (i.e., hand delivered)

_____ Certified Mail   Certified Receipt Received: Yes [ ]   No [ ]

_____ First-Class Mail

_____ Posting of notice in conspicuous place in or about the structure

Photograph of posting   Yes [ ]   No [ ]

(Continued)

Other inspections or contacts:

Date: _____   Summary of findings: _____

New evidence: _____
Yes [ ]   No [ ]   _____

                     _____

Date: _____   Summary of findings: _____

New evidence: _____
Yes [ ]   No [ ]   _____

                     _____

Date: _____   Summary of findings: _____

New evidence: _____
Yes [ ]   No [ ]   _____

                     _____

Date: _____   Summary of findings: _____

New evidence: _____
Yes [ ]   No [ ]   _____

                     _____

Date: _____   Signature of Inspector: _____

SAMPLE FORM

REINSPECTION RECORD—IPMC 104.3

Location of Property: _____

PIN: _____

Problem(s) corrected: [ ] Yes   [ ] No

Inspection Date: _____ Time: _____

Summary of Findings: _____

_____

_____

_____

Name of continuing violation(s)   Code Section(s) violated:

_____   _____

_____   _____

_____   _____

Photos: Yes [ ]   No [ ]   Video: Yes [ ]   No [ ]

Contact with owner: Yes [ ]   No [ ]

If yes, summary of conversation _____

_____

_____

If property not in compliance:

[ ] Referred to prosecutor   Date: _____

[ ] Complaint(s) prepared for court Date: _____

[ ] Complaints filed in court   Date: _____

[ ] Set for administrative hearing   Date: _____

[ ] Referred to social services   Date: _____

[ ] Consultation with supervisor   Date: _____

[ ] No action taken

[ ] Other _____

Date: _____   Signature of Inspector: _____

SAMPLE FORM
- -

VILLAGE OF WOODRIDGE MULTI-FAMILY HOUSING INSPECTION FORM

**NOTE: The items on this list are items that are typically looked at during the normal inspection procedure. Although every effort has been made to make this list as complete as possible, it is impossible to include every situation or circumstance, therefor, the items noted on the inspection reports will not be limited to this list.

INTERIOR DWELLING UNITS:

A) LIVING ROOM, DINING ROOM AND HALLWAY AREAS;

1) Are smoke and heat detectors in place and operating properly? Have the detectors been replaced, if so, were they replaced with the proper approved device as required by fire code?

    Comment:_____

2) Are electric fixtures such as air conditioners, lights, fans etc. installed properly and are they in good working condition?

    Comment:_____

3) Is there power to all the receptacles in the unit, is the polarity correct (are they wired correctly), this can be tested with a standard G.F.I. Tester, are they filled with paint or are they cracked or damaged in any way?

    Comment:_____

4) Do all the light switches work, are they cracked or do you notice any arcing or sparks in the switch when operating it?

    Comment:_____

5) Are any of the receptacles or light switches missing cover plates?

    Comment:_____

6) If there is a circuit breaker panel in the unit, is it clearly and correctly marked or labelled? Are there any open knockouts that need to be plugged?

    Comment:_____

7) Are there any extension cords or cable wires running through any windows or doorways?

    Comment:_____

8) Are the entry doors and jambs to the unit in good condition, do they lock and latch properly, do they have the proper security locks and door scopes as required by Security Code?

    Comment:_____

9) Does the patio door operate properly, does it latch and lock, is the seal around the glass in good condition or is it bad and causing the door to cloud up?

    Comment:_____

(Continued)

VILLAGE OF WOODRIDGE MULTI-FAMILY HOUSING INSPECTION FORM (*CONTINUED*)

10) Is the patio door screen in place and operating properly, is it damaged in any way?
Comment:_____

11) Are the windows in good weather tight condition, do they have bad seals causing them to cloud up, do they operate properly, is there any cracked or broken glass, are all the window screens in place and are they damaged in any way?
Comment:_____

12) Do the closet doors operate properly, are they damaged in any way, is the closet shelving in place and secure?
Comment:_____

13) Is there any damage to the walls or ceilings such as holes, cracks, water stains, drywall nails popping out, peeling paint etc.?
Comment:_____

14) Are the floors and floor coverings in good condition, is there any wrinkled or torn carpeting that can cause a trip hazard?
Comment:_____

15) Does the intercom and door buzzer work properly?
Comment:_____

B) KITCHEN;

1) Do the plumbing fixtures operate properly including supply valves, are there any leaks?
Comment:_____

2) Are the electric fixtures such as hood fans, lights, exhaust fans etc. installed and operating properly, are the receptacles and switches in good condition?
Comment:_____

3) Is the oven in good working condition, is the gas hooked up correctly, is there a smell of gas in the room, does the oven door close completely, are all the control knobs in place and operating properly?
Comment:_____

4) Is the refrigerator and freezer in good working condition, do the doors close and seal properly?
Comment:_____

5) If there is a garbage disposal and/or dishwasher, are they in good working order?
Comment:_____

6) Are the cabinets and counter tops in good condition?
Comment:_____

7) Are the walls and Ceiling in good condition, is the floor tile in good condition?
Comment:_____

(Continued)

C) BATHROOM;

1) Are plumbing fixtures in place and operating properly, including toilet, tub and shower units, sinks, supply valves, etc., are there any leaks?

   Comment:_____

2) Are the electrical fixtures, including lights, exhaust fans, etc., installed and operating properly?

   Comment:_____

3) Is the G.F.I. receptacle wired and functioning correctly, is the light switch operating properly? If there is not a G.F.I. outlet in this room, when or if the existing outlet is replaced it must be replaced with a G.F.I. outlet.

   Comment:_____

4) If there is a window, is it in good weather tight condition, is there a screen in the window, is there any cracked or broken glass?

   Comment:_____

5) Are the walls, ceiling and floor in good condition, is there any pealing paint, water stains, loose or missing tile, is there any mold, etc.?

   Comment:_____

D) BEDROOMS, DENS AND OTHER LIVING AREAS;

1) Are all electrical fixtures installed and operating properly?

   Comment:_____

2) Is there power to all receptacles, are they filled with paint or cracked, are the switches in good working order, are all the cover plates in place?

   Comment:_____

3) Are the closet doors and room doors and jambs in good condition and functioning properly, are the closet shelves in good condition and secure?

   Comment:_____

4) Are the windows in good weather tight condition, do they function properly, are the screens in place and in good condition, is there any broken glass?

   Comment:_____

5) Are the walls, ceiling and floor in good condition, are there any water stains, pealing paint, holes, cracks, nail pops, are there any trip hazards such as torn or wrinkled carpet?

   Comment:_____

E) GENERAL;

1) Are the balcony railings secure, is the railing missing any balusters, is it rusty or in need of paint, is the balcony in good condition, is it in need of paint?

   Comment:_____

2) Is the unit reasonably clean, is there a bug problem?

   Comment:_____

(Continued)

VILLAGE OF WOODRIDGE MULTI-FAMILY HOUSING INSPECTION FORM (*CONTINUED*)

INTERIOR COMMONS:

F) FURNACE/BOILER ROOM;

1) Is the heat detector in place and operating, if the room is sprinklered, is the sprinkler head in good condition, is it clean?

   Comment:_____

2) Is there any penetrations through the ceiling, walls, or floor?

   Comment:_____

3) Is there any storage in the room, is it combustible or flammable, if it is not combustible or flammable, is it blocking access to any of the heating, water heater or other units?

   Comment:_____

4) Are the pressure discharge pipes from the water heater and boiler made of rigid metal, are they directed in such a way as not to cause a potential hazard?

   Comment:_____

5) Is the fire alarm panel in this room, is it working properly? Is the trouble indicator light on, if so this usually indicates a problem, call your service company.

   Comment:_____

6) Are there any open electrical junction boxes, are there any open knockouts that need to be plugged in any of the electrical junction boxes or panels?

   Comment:_____

7) Is the door and jamb in good condition, is it self closing and latching, does the door have the proper fire rating?

   Comment:_____

8) Is there a working light fixture in the room, is it properly installed and in good condition?

   Comment:_____

9) Is the State Boiler Inspection Certificate posted in the room, is it up to date?

   Comment:_____

10) If there is a circuit breaker panel in the room, is it labeled correctly, is there any knock-outs that need to be plugged?

   Comment:_____

11) If there is a window in the room, is it in good weather tight condition, is there any broken glass etc.?

   Comment:_____

12) Are there any plumbing leaks?

   Comment:_____

13) If there is a floor drain in the room, is the grate or cover in place?

   Comment:_____

14) Is all the venting properly installed and in good condition?

   Comment:_____

(Continued)

G) LAUNDRY ROOM;

1) Is the heat detector in place and operating, if the room is sprinklered, is the sprinkler head in good condition, is it clean?

   Comment:_____

2) Is the door self closing and latching?

   Comment:_____

3) Is the fire alarm panel in this room, is it working properly? Is the trouble indicator light on, if so this usually means there is a problem, call your service company.

   Comment:_____

4) Is there any storage in the room, is it combustible or flammable, if it is not combustible or flammable, is it blocking access to the door or any of the appliances?

   Comment:_____

5) Are there any penetrations in the walls, ceiling or floor?

   Comment:_____

6) Is the pressure discharge pipe from the water heater made of rigid metal, is it directed in such a way as not to cause a potential hazard?

   Comment:_____

7) Are there any open electrical junction boxes, are there any open knockouts that need to be plugged?

   Comment:_____

8) Is there a circuit breaker panel in this room, is it properly labeled, is there any open knockouts that need to be plugged?

   Comment:_____

9) Is there a light fixture in the room, is it in good condition, is it installed and working properly?

   Comment:_____

10) Are there any plumbing leaks?

   Comment:_____

11) Is the dryer exhaust connected and vented properly, is the damper to the dryer vant clean and functioning properly?

   Comment:_____

12) If there is a floor drain in this room, is the grate or cover in place?

   Comment:_____

13) If the room has an exhaust fan, is it in good condition and operating properly, are all the switches and receptacles in good condition and are all the cover plates in place?

   Comment:_____

(Continued)

VILLAGE OF WOODRIDGE MULTI-FAMILY HOUSING INSPECTION FORM (*CONTINUED*)

14) Is the floor tile in good condition, are there any loose tiles etc.?

Comment:_____

15) Are the walls, ceiling, doors and jambs in good condition, are they in need of paint, is there any holes, cracks or water stains etc.?

Comment:_____

H) ELECTRIC/METER ROOM;

1) Is the heat detector in place and operating, has the detector been replaced, if so, was it replaced with the proper approved device as required by Fire Code?

Comment:_____

2) Is the fire alarm panel located in this room, is it working properly? Is the trouble indicator light on, if so, this usually indicates a problem, call your service company.

Comment:_____

3) Is there any combustible or flammable storage in the room, is there any storage that is blocking access to doors, panels or any other electrical equipment?

Comment:_____

4) Is there any penetrations in the walls ceiling or floor?

Comment:_____

5) Are all the circuit breakers labeled correctly, is there any knockouts that need to be plugged?

Comment:_____

6) Is there a window in this room, if so, is it in good weather tight condition, is there any broken glass?

Comment:_____

7) Are there any open electrical junction boxes, are all the cover panels to the electrical equipment secure?

Comment:_____

8) Are the walls and ceiling in this room in good condition, are there any cracks, holes, water stains etc.?

Comment:_____

I) HALLWAYS AND STAIRWAYS;

1) Are the emergency and exit lights in good working condition?

Comment:_____

2) Are the fire extinguishers in place and up to date, are the fire extinguisher cabinets in good condition, is the glass missing or broken?

Comment:_____

3) Are the fire alarm pull stations in place and in good condition, are the glass bars missing?

Comment:_____

(Continued)

4) Are the smoke and heat detectors in place and operating, have they been replaced, if so, were they replaced with the proper approved device as required by Fire Code?

Comment:_____

5) Is there any storage in these areas, that is blocking any part of the means of egress?

Comment:_____

6) Is the carpet and tile in good condition, is there any possible trip hazards?

Comment:_____

7) Are the railings on the stairs and landings in good condition, are they secure, are there any missing balusters?

Comment:_____

8) Are the smoke and fire doors and jambs in good working condition, do they self close and self latch?

Comment:_____

9) Are the entry and security doors and jambs in good condition, do they self close, self latch and do the locking systems work properly, are they weather tight?

Comment:_____

10) Are the light fixtures in good working condition, are there any missing parts?

Comment:_____

11) Are the receptacles and light switches in good condition, are they filled with paint or cracked, are all the cover plates in place?

Comment:_____

12) Are there any windows in this area, are they in good weather tight condition, is there any broken glass, are the screens in place, are the screens in good condition?

Comment:_____

13) Are the walls and ceiling in this area in good condition, is there any cracks, holes, water stains etc.?

Comment:_____

14) Is this area reasonably clean, is there any sign of a bug problem?

Comment:_____

EXTERIOR COMMONS:

J) EXTERIOR STRUCTURE;

1) Are all exterior walls, soffit and facia etc., free of holes, breaks, loose or rotting materials and maintained weatherproof and properly surface coated to prevent deterioration?

Comment:_____

2) Is any of the masonry in need of repair or tuckpointing?

Comment:_____

3) Are all the gutters and downspouts in good working condition and free of debris?

Comment:_____

(Continued)

VILLAGE OF WOODRIDGE MULTI-FAMILY HOUSING INSPECTION FORM (*CONTINUED*)

4) Are the downspout extensions or splashblocks in their proper place as to direct roof drainage away from building foundation?

Comment:_____

5) Are all exterior electrical installations properly installed with materials approved for such use?

Comment:_____

6) Are any of the exterior light fixtures missing any parts, are they in good working order?

Comment:_____

7) Is any of the concrete in need of repair or replacement, are any of the stoops or patios sinking, remember the maximum rise for a step is 7 inches.

Comment:_____

8) Is the roofing and flashing in good condition and free of defects that might admit rain?

Comment:_____

9) Are exterior stairway handrails in place and secure?

Comment:_____

10) Are the building address numbers in good condition and clearly visible from the street or parking lot?

Comment:_____

K) EXTERIOR LANDSCAPING AND GROUNDS;

1) Are there any trip hazards along the sidewalks or stoops?

Comment:_____

2) Are all parking lot and driveway areas in a proper state of repair and maintained free of hazardous conditions?

Comment:_____

3) Is signage properly installed and maintained?

Comment:_____

4) Are dumpster areas reasonably clean, are dumpster screens in good condition?

Comment:_____

5) Are there any grading problems, standing water?

Comment:_____

6) Are there any muddy areas in need of sod or grass?

Comment:_____

7) Is there any dead landscaping in need of replacement?

Comment:_____

8) Is there any land erosion under stoops; patios or sidewalks?

Comment:_____

(Continued)

9) Are there any abandoned cars on the property?

Comment:_____

10) Does the exterior lighting illuminate all areas of the walkways and parking areas sufficiently?

Comment:_____

FLOWCHART 1

BASIC ENFORCEMENT PROCEDURE (PERMISSION GIVEN FOR INSPECTION OR NO PERMISSION NEEDED)

COMPLAINT RECEIVED OR OBSERVED BY STAFF
(USE SAMPLE FORM: IPMC 104.3—INSPECTION RECORD)

↓

INSPECTION PERFORMED
(USE SAMPLE FORM: IPMC 104.3—INSPECTION RECORD)

↓

PREPARE NOTICE OF VIOLATION
(USE CHECKLIST: IPMC 107.1—LEGAL RESPONSIBILITY AND NOTICE; SAMPLE FORMS: RESPONSE TO RESPONSIBLE PARTY FOR EXTENSION OF TIME OR NOTICE TO RESPONSIBLE PARTY—IPMC 107.2 AND CHECKLIST: SENDING NOTICE OF VIOLATION)
OR CASE CLOSED BECAUSE NO VIOLATION FOUND

↓

REINSPECTION OF PROPERTY AFTER COMPLIANCE DATE
(USE SAMPLE FORM: IPMC 104.3—REINSPECTION RECORD)

↓

PREPARATION OF COMPLAINT
(USE CHECKLIST: IPMC 106—DRAFTING THE COMPLAINT AND SELECT APPROPRIATE SAMPLE COMPLAINT)
OR CASE CLOSED BECAUSE OF COMPLIANCE

↓

SUBMIT COMPLAINT TO COURT OR TO ADMINISTRATIVE HEARING BODY

↓

MAKE SURE DEFENDANT IS SERVED WITH COMPLAINT ACCORDING TO RULES OF LOCAL JURISDICTION

↓

GO TO COURT OR ADMINISTRATIVE HEARING

BASIC ENFORCEMENT PROCEDURE (NO PERMISSION FOR INSPECTION)

COMPLAINT RECEIVED OR OBSERVED BY STAFF
(USE SAMPLE FORM: IPMC 104.3—INSPECTION RECORD)

↓

INSPECTION NOT PERFORMED BECAUSE PERMISSION REFUSED
(USE CHECKLIST: RIGHT OF ENTRY—IPMC 104.4)

↓

DRAFT ADMINISTRATIVE SEARCH WARRANT
(USE FORMS: COMPLAINT FOR ADMINISTRATIVE SEARCH WARRANT, INSPECTION LIST,
ADMINISTRATIVE SEARCH WARRANT, SEARCH WARRANT RETURN —IPMC 104.4

↓

EXECUTE ADMINISTRATIVE SEARCH WARRANT AND PERFORM INSPECTION
(USE SAMPLE FORM: IPMC 104.3—INSPECTION RECORD)

↓

PREPARE NOTICE OF VIOLATION
(USE CHECKLIST: IPMC 107.1—LEGAL RESPONSIBILITY AND NOTICE and SAMPLE FORM:
NOTICE TO RESPONSIBLE PARTY—IPMC 107.2)
OR CASE CLOSED BECAUSE NO VIOLATION FOUND

↓

**REINSPECTION OF PROPERTY AFTER COMPLIANCE DATE (IF PERMISSION
REFUSED, SEEK NEW ADMINISTRATIVE SEARCH WARRANT**
(USE SAMPLE FORM: IPMC 104.3—REINSPECTION RECORD)

↓

PREPARATION OF COMPLAINT
(USE CHECKLIST: IPMC 106—DRAFTING THE COMPLAINT AND SELECT APPROPRIATE
SAMPLE COMPLAINT)
OR CASE CLOSED BECAUSE OF COMPLIANCE

↓

SUBMIT COMPLAINT TO COURT OR TO ADMINISTRATIVE HEARING BODY

↓

**MAKE SURE DEFENDANT IS SERVED WITH COMPLAINT ACCORDING TO RULES
OF LOCAL JURISDICTION**

↓

GO TO COURT OR ADMINISTRATIVE HEARING

CHECKLIST

RIGHT OF ENTRY—IPMC 104.4

❑ Can the violation be seen from public property or from someone else's private property?

❑ Will the owner or occupant consent to the inspection?

❑ Will the agent or management company representing the owner consent to the inspection?

❑ Are there other persons having control over the property who can consent to an inspection?

❑ Has the owner or occupant refused entry?

❑ If entry has been refused, what action should be taken next? Gather more evidence? Seek legal counsel?

❑ Are there exigent circumstances that allow entry?

❑ If commercial property, is the business a closely regulated one that is subject to an inspection without a warrant?

❑ Is there enough information to show probable cause that there is a violation on the premises?

❑ Should an administrative search warrant be obtained?

❑ Has everyone who has knowledge about the condition of the premises been interviewed?

❑ Should other agencies (e.g., health department, fire department) be involved in the inspection?

❑ Is this a situation where a regular inspection is essential for the health and safety of the surrounding community and the occupants?

❑ Has legal counsel been consulted about obtaining the warrant or has counsel reviewed the document?

SAMPLE FORM

COMPLAINT FOR ADMINISTRATIVE SEARCH WARRANT—IPMC 104.4

IN THE MATTER OF

1020 Peal Road
Woodridge, Illinois

CASE NUMBER: *06 MR 0001*

COMPLAINT FOR ADMINISTRATIVE SEARCH WARRANT

NOW APPEARS *Karyn Byrne*, Code Enforcement Officer of the *Village of Woodridge*, Complainant, before the undersigned Judge of the *18th Judicial Circuit*, and requests the issuance of an Administrative Search Warrant, to inspect the premises of *1020 Peal Road, Woodridge, Illinois*, to determine if said premises are maintained in compliance with the ordinances of the *Village of Woodridge*, being the *2006* International Property Maintenance Code of 2006, as adopted and amended in Section *8-1J-1(A) of the Village of Woodridge Code*. In support hereof, Complainant states as follows:

1. The *2006* International Property Maintenance Code as amended and adopted by reference in Section *8-1J-1(A)* of the *Village of Woodridge* Code prescribes minimum maintenance standards for all structures and premises for basic equipment and facilities for light, ventilation, occupancy limits, heating, plumbing, electricity, and sanitation; for safety from fire; for space, use, and location; and for safe and sanitary maintenance for all structures and premises now in existence.

2. Section IPMC 104.4 of the *2006* International Property Maintenance Code as amended and adopted by reference in Section *8-1J-1(A)* of the *Village of Woodridge* Code provides that: "The code official is authorized to enter the structure or premises at reasonable times to inspect subject to constitutional restrictions on unreasonable searches and seizures. If entry is refused or not obtained, the code official is authorized to pursue recourse as provided by law."*

3. The Complainant,** if allowed entry into the premises, shall inspect the items listed on the attached inspection form, which is attached hereto and made a part hereof.

4. There is probable cause for the administrative search warrant based on the following facts:***

    a. *On June 22, 2006 at 9:00 a.m. I had a conversation with Officer Steven Herron, a police officer for the Village of Woodridge. He called me asking for assistance because he had been called to 1020 Peal Road, Woodridge, Illinois, based on a call for help to 911. When he arrived with the paramedics, he discovered an elderly couple, Joseph and Lauren McKenna. Mr. McKenna was having trouble breathing and was attended by the paramedics. Officer Herron noticed that the house was not in a safe or sanitary condition. There were boxes and newspapers piled up to the ceiling in the living room. He observed moldy dishes in the kitchen sink and what appeared to be cat feces on the floor in each room. He observed eight cats while on the premises. Officer Herron asked that I inspect the property for property maintenance violations.*

    b. *On June 22, 2006 at 3 p.m. I went to 1020 Peal Road, Woodridge, Illinois, and observed the property from the public sidewalk. I observed that the roof was missing a number of shingles and that there was peeling paint on the structure. I also noticed that the front window of the home was blocked by what appeared to be piles of newspapers. I made contact with Mrs. Lauren*

(Continued)

McKenna by knocking on the door, the doorbell being inoperative. Mrs. McKenna appeared at the front door and I presented my identification from the Village of Woodridge to her. I told her I was concerned about her safety and that of her husband and that I would like to inspect the residence and assist her if I found anything hazardous. Mrs. McKenna told me that it was not a convenient time and she would call me when it was. While I was standing at the door, I could smell an offensive odor believed to be cat urine and noticed 3 cats. I gave her my card and departed the premises.

    c. On June 23, 26, and 27, 2006 I called the number for Mr. and Mrs. Joseph McKenna but no one answered. No one called me regarding scheduling an inspection. On June 28, 2006 at 10 a.m. I called the number again and spoke with a man identifying himself as Mr. Joseph McKenna. He told me that he wanted me "to leave them alone and let them live the way they want to without interference from nosy people."

5.   Complainant reasonably believes that the property at *1020 Peal Road, Woodridge, Illinois,* is in violation of the following sections of the *2006* International Property Maintenance Code as amended and adopted by reference in Section *8-1J-1(A)* of the *Village of Woodridge* Code: *Structure Unfit for Human Occupancy, 108.1.3, Failure to Maintain Exterior Surfaces, 304.2, Failure to Maintain Roof, 304.7, Failure to Maintain Interior Structure in Sanitary Condition, 305.1, and Accumulation of Rubbish or Garbage, 307.1.*

WHEREFORE, Complainant prays that this Court issue an Administrative Search Warrant, to inspect the structure and property at *1020 Peal Road, Woodridge, Illinois.*

Karyn Byrne
Complainant

Subscribed and sworn to before me
on this *28th* day of *June, 2006*

_____
Notary Public or Judge

  *If there is an additional specific state statute or municipal code section that authorizes an administrative search warrant include the following line:
    "Section _____ of the (_____ state statute) (_____ municipal code) provides for the issuance of an administrative search warrant."

 **List anyone else who may accompany the inspector, e.g., Animal Control, Health Department, a police officer, for the limited purpose of providing security, or building code inspectors.

***Set forth all the facts that form the probable cause for the search warrant or, if probable cause does not exist, the basis for an administrative inspection.

COMMENT: If there is a state statute that sets forth the forms to be used for obtaining an administrative search warrant, the code official should use those forms and should follow the statutory procedure. The inspector should always seek legal assistance in drafting and obtaining an administrative search warrant to make sure that all the statutory and constitutional requirements are satisfied.

INSPECTION LIST—IPMC 104.4

Check all that apply:

❑ Structural members will be checked for any evidence of deterioration that would render them incapable of carrying the imposed loads.

❑ The exterior of the property and premises will be inspected to determine that it is in a clean, safe, and sanitary condition free from the accumulation of rubbish or garbage and to ensure that the exterior structure is in good repair and structurally sound.

❑ The interior of the structure and its equipment will be examined to make sure it is in good repair, structurally sound, and in a sanitary condition, so as not to pose a threat to the health, safety, or welfare of the occupants or visitors, and to protect the occupants from the environment.

❑ The structure will be checked for dampness that would be conducive to decay or deterioration of the structure.

❑ The sanitation of the structure will be inspected to ensure that it is in a clean and sanitary condition free from any accumulation of rubbish or garbage.

❑ The plumbing facilities will be inspected to ensure that they are in proper operating conditions, can be used in privacy, and are adequate for personal cleanliness and the disposal of human waste. The plumbing fixtures will be examined to ensure that they are maintained in a safe and usable condition, and to make sure that they are of approved material. The sink, lavatory, bathtub or shower, water closet, or other facility will be inspected to ensure that they are properly connected to either a public water system or to an approved private water system.

❑ All mechanical equipment will be checked to ensure that it is properly installed and safely maintained in good working condition, and that it is capable of performing the function for which it was designed and intended.

❑ All electrical equipment, wiring, and appliances shall be inspected to ensure that they are installed and maintained in a safe manner.

❑ The residences shall be inspected for any signs of insects, rats, or other pests that could require extermination.

[If the search warrant is for over-occupancy, include the following:]

❑ All places where persons might sleep, including rooms and closets, will be inspected for any signs of over–occupancy, and all drawers that might contain evidence of how many persons live in the structure, such as mail and identifying material, will be inspected.

COMMENT: If the property maintenance inspector does not have the necessary qualifications, a building code inspector or other experts should be part of the inspection team if some parts of the inspection go beyond the inspector's knowledge, e.g., inspecting structural members. Their presence should be noted in the complaint for administrative search warrant, paragraph 3.

ADMINISTRATIVE SEARCH WARRANT—IPMC 104.4

IN THE MATTER OF

1020 Peal Road
Woodridge, Illinois

CASE NUMBER: *06 MR 0001*

ADMINISTRATIVE SEARCH WARRANT

On this day, *June 28, 2006 at 3:00 p.m.,* Complainant, *Karyn Byrne,* has subscribed and sworn to a Complaint for an Administrative Search Warrant before me. Upon examination of the Complaint, I find that it states facts to show a reasonable basis and probable cause, and I therefore command that the structure and property at *1020 Peal Road, Woodridge, Illinois,* be inspected as set forth in the inspection list attached hereto and made a part hereof, including the interior and all rooms therein and exterior of the structure,* and videotaped and/or photographed to determine if said premises are in compliance with the *2006* International Property Maintenance Code as amended and adopted by reference in Section *8-1J-1(A)* of the *Village of Woodridge* Code, specifically sections** *IPMC 108.1.3, 304.2, 304.7, 305.1, and 307.1.*

Time and date of Issuance:
June 28, 2006, at 3:00 p.m.

_____
Judge or Magistrate

  *List specific areas to be searched, especially closets and drawers if appropriate.

**List specific ordinance sections covering those violations the inspector believes will be found.

SEARCH WARRANT RETURN—IPMC 104.4

IN THE MATTER OF

1020 Peal Road
Woodridge, Illinois

CASE NUMBER: *06 MR 0001*

Premises: *1020 Peal Road, Woodridge, Illinois*

I served this Administrative Search Warrant on the above described Premises and aided in its execution, on the *28th* day of *June, 2006*, at *4:00* p.m.

Date: *June 29, 2006*

Karyn Byrne

Complainant

Or:

I did not serve this Administrative Search Warrant within *96** hours of the time of issuance, and it is hereby returned to the Court as void and not executed.

Date: *July 3, 2006*

Karyn Byrne

Complainant

* Insert the time limit set by state statute for search warrant execution.

IPMC 106 Enforcement Procedures and Administrative Violations

CHECKLIST

DRAFTING THE COMPLAINT—IPMC 106

❑ Is the name on the complaint proper?

❑ Is the address a place where the defendant can be reached, such as an address from the tax records?

❑ Is the date and time on the complaint the same as when the inspection occurred?

❑ Are the right offense and section number cited? Should the defendant be cited for multiple charges?

❑ Is the description of the offense adequate so that the defendant can defend himself or herself?

❑ Is the complaint signed?

❑ Is the complaint notarized? (If required by state law)

❑ Has the charge been properly served?

❑ When is the court date?

❑ Has the property been inspected before the court date so that the judge knows whether there is compliance?

❑ Does the inspector have identifying information on the defendant, e.g., date of birth, height, weight, driver's license number?

❑ Can the inspector testify in court as to the identity of the defendant?

SAMPLE COMPLAINT

FAILING TO COMPLY WITH A NOTICE OF VIOLATION—IPMC 106.3

STATE OF *ILLINOIS*
COUNTY OF DUPAGE
VILLAGE OF WOODRIDGE
v.
NAME: *JOHN FINCHAM*
ADDRESS: *6560 Hollywood Blvd.*
CITY: *Woodridge, IL 60517*

The undersigned says that on or about *April 28, 2006*, at or about *3:00 p.m.*, the Defendant did unlawfully commit the offense of **Failing To Comply with a Notice of Violation** in violation of ***2006 IPMC 106.3*** as amended and adopted by reference in Section *8-1J-1(A)* of the *Village of Woodridge* Code; in that said Defendant, being the owner* of *6560 Hollywood Blvd., Woodridge, IL*, did fail to comply with a notice of violation sent by code official, *Karyn Byrne*, in that said Defendant failed to correct the violation on the property at *6560 Hollywood Blvd., Woodridge, IL*, as ordered by *Karyn Byrne*, a code official for the *Village of Woodridge*, as set forth in the notice of violation on *April 3, 2006*, a copy of which is attached hereto and made a part hereof.

Karyn Byrne

Code Official

Sworn to and Subscribed before Me
This 1st Day of August, 2006

_____

Notary Public

*Or, occupant

IPMC 107   Notices and Orders

CHECKLIST

LEGAL RESPONSIBILITY AND NOTICE—IPMC 107.1

I. DETERMINING LEGAL RESPONSIBILITY

❑ Does the code set forth who is legally responsible for the violation?

❑ Has the responsible party been determined?

❑ Is there an admission from the responsible party?

❑ Have the records of the Recorder of Deeds been checked to verify who owns the property?

❑ Is there more than one owner of the property?

❑ If it is believed that the responsible party is a corporation, or limited liability company, has that information been verified with the Secretary of State?

❑ If the corporation in good standing or has it been dissolved?

❑ Have all possible records been checked for information on the responsible party? Recorder of Deeds? Permits? Business licenses?

❑ Should a trust disclosure letter be done to find out who the beneficiaries of a land trust are?

II. CONTENT OF NOTICE OF VIOLATION

❑ Has the notice been reduced to writing?

❑ Is the description of the real estate sufficient for identification, e.g., address or property index number?

❑ If the property is vacant land, does the inspector have the property index number?

❑ Does the notice contain the statement of violation and why the notice has been issued i.e., name, description, and section number of the violation?

❑ Has the responsible party been given a correction order allowing a reasonable amount of time to make the repairs and improvements required to bring the dwelling unit or structure into compliance with the provisions of the code?

❑ Has the responsible party obtained the necessary building permits to perform the work?

❑ Has the property owner been notified about the right to appeal?

❑ Is a statement of the right of the municipality to file a lien included?

❑ Has the inspector checked with legal counsel about the correct way to cite the code in the notice of violation?

NOTICE TO RESPONSIBLE PARTY—IPMC 107.2

April 3, 2006

Mr. John Fincham
6560 Hollywood Blvd.
Woodridge, IL 60517

Re: *6560 Hollywood Blvd., Woodridge, Illinois*
Property Index Number: *01-0001-001-00*

Dear *Mr. Fincham*:

An inspection of your property at *6560 Hollywood Blvd. Woodridge, Illinois,* on *April 2, 2006,* shows the following violations of the code of ordinances of the *Village of Woodridge:*

Accumulation of Rubbish in violation of 2006 IPMC-307.1 as amended and adopted by reference in Section *8-1J-1(A)* of the *Village of Woodridge* Code—*the backyard of the premises has an accumulation of tires, containers, automobile parts, and rusted equipment. The backyard must be free of any accumulation of rubbish.*

Failure of Habitable Space to Have Openable Window in violation of 2006 IPMC 403.1 as amended and adopted by reference in Section *8-1J-1(A)* of the *Village of Woodridge* Code—*no windows in the master bedroom can be opened because they are painted shut. Every habitable space must have at least one openable window.*

A correction of these problems must be made by the close of business on *April 22, 2006,* or a complaint will be filed against you in a court of local jurisdiction.

You have a right to appeal this notice and order by filing a written application for appeal with the Board of Appeals for the *Village of Woodridge.* The application for appeal must be filed within twenty (20) days after the day this notice is served upon you. The appeal shall be based on a claim that the true intent of the code or the rules legally adopted thereunder have been incorrectly interpreted, the provisions of the code do not fully apply, or the requirements of the code are adequately satisfied by other means.

If you fail to correct these violations, any action taken by the *Village of Woodridge,* the authority having jurisdiction, may be charged against the real estate upon which the structure is located and shall be a lien upon such real estate.

Please feel free to contact me to discuss this matter further.

Very truly yours

Karyn Byrne

Deputy Code Official

COMMENT: The Board of Appeals language in Section 111.2 and the length of time given for appeals in Section 111.1 are the most frequently amended sections of the 2006 International Property Maintenance Code. They often differ from the model code sections. Therefore, the inspector should make sure this form is adapted to conform to the local jurisdiction's requirements.

PROOF OF SERVICE OF NOTICE—IPMC 107.2

PROOF OF SERVICE

To:* Name: *Mr. John Fincham*
　　　Last known address: *6560 Hollywood Blvd.*
　　　　　　　　　　　　　Woodridge, IL 60517

On *April 3, 2006*, I, *Karyn Byrne*, on oath state that:**

❏ I served this notice by delivering a copy personally to each person to whom it is directed.
　　　　　　　　　　　　Signature of recipient_____

❏ I served this notice by mailing a copy to each person to whom it is directed and depositing the same in the U.S. Mail at Seven Plaza Drive, Woodridge, Illinois, with the proper first-class postage prepaid.

❏ I served this notice by mailing a copy to each person to whom it is directed by certified mail and depositing the same in the U.S. Mail at Seven Plaza Drive, Woodridge, Illinois, with the proper postage prepaid. Certified mail number _____.

❏ I served this notice by posting it in a conspicuous place on or about the structure affected by the notice, being *6560 Hollywood Blvd., Woodridge, IL.*

　　　　　　　Photograph of posting taken?   [ ] Yes   [ ] No

Subscribed and sworn to before me this:

Date: *April 3, 2006*

　　　　　　　　　　　　Karyn Byrne

　　　　　　　　　　　　Deputy Code Official

_____

Notary Public

 *List the name and addresses of all parties to whom the notice is directed.
**Check the appropriate box.

COMMENT: When serving the notice by posting it, the inspector should take a picture of the notice on the building so the defendant does not claim it was never there.

RESPONSE TO RESPONSIBLE PARTY FOR EXTENSION OF TIME—IPMC 107.2

April 6, 2006

Mr. John Fincham
6560 Hollywood Blvd.
Woodridge, IL 60517

Re: *6560 Hollywood Blvd., Woodridge, Illinois*
Property Index Number: *01-0001-001-00*

Dear *Mr. Fincham*:

Thank you for contacting me so promptly regarding the notice of violation I sent you on behalf of the *Village of Woodridge* on *April 3, 2006*. Based on your request, I am extending the deadline for correction of the violations from *April 12, 2006* to the close of business on *April 24, 2006*. Please contact me by *April 24, 2006* to schedule an inspection so I may verify that the corrections have been completed.

Very truly yours

Karyn Byrne

Deputy Code Official

CHECKLIST

SENDING NOTICE OF VIOLATION—IPMC 107.2

❑ What does the code say about the content of the notice of violation and the way it has to be served?

❑ Has the responsible party or parties been given notice?

❑ If the responsible party is a corporation or limited liability company, has the inspector found out who the registered agent is for the legal entity?

❑ If the responsible party is a corporation, has the inspector found out who the officers or directors are?

❑ What proof of service exists?

❑ If served by mail, was a proof of service form prepared?

❑ Should the responsible party be given an extension of time to correct the violation? What is reasonable?

❑ How was service made:

    ❑ Delivered personally

    ❑ Sent by certified mail addressed to the last known address

    ❑ Sent by first-class mail addressed to the last known address

    ❑ Posted in a conspicuous place in or about the structure affected by the notice.

❑ Did the recipient of the notice sign a receipt for the notice?

❑ If service was made by posting, was a photograph taken to prove that it was served in this manner?

UNLAWFUL TRANSFER OF OWNERSHIP—IPMC 107.5

STATE OF *ILLINOIS*
COUNTY OF DUPAGE
VILLAGE OF WOODRIDGE
v.
NAME: *JOHN FINCHAM*
ADDRESS: *6560 Hollywood Blvd.*
CITY: *Woodridge, IL 60517*

The undersigned says that on or about *May 24, 2006,* the Defendant did unlawfully commit the offense of **Unlawful Transfer of Ownership** in violation of ***2006* IPMC 107.5** as amended and adopted by reference in Section *8-1J-1(A)* of the *Village of Woodridge* Code; in that said Defendant, being the owner *of 6560 Hollywood Blvd., Woodridge, IL,* did unlawfully sell* said dwelling unit** to *Tim Halik* after receiving a compliance order dated *April 3, 2006,* which is attached hereto and made a part hereof, from the *Village of Woodridge* prior to complying with the order or notice of violation*** sent by code official, *Karyn Byrne.*

Karyn Byrne

Deputy Code Official

Sworn to and Subscribed before Me
This *1st Day* of *August, 2006*

_____
Notary Public

  *Or, (choose one) transfer, mortgage, lease or otherwise dispose of.

 **Or, structure.

***Or, prior to furnishing the (choose one) grantee, transferee, mortgagee, or lessee a true copy of the compliance order or notice of violation issued by the code official and/or prior to furnishing the code official a signed and notarized statement from the (choose one) grantee, transferee, mortgagee, or lessee acknowledging the receipt of the compliance order or notice of violation and fully accepting unconditional responsibility for making the corrections or repairs required by the compliance order or notice of violation.

IPMC 108　Unsafe Structures and Equipment

IPMC 108.5—Prohibited Occupancy

CHECKLIST

UNSAFE STRUCTURES AND EQUIPMENT

❑ Does the inspector possess the necessary expertise to find that the structure or equipment is unsafe, unfit for human occupancy, or unlawful?

❑ Should the inspector obtain the services of an expert before finding the structure or equipment unsafe or unfit for human occupancy?

❑ Should the inspector seek legal advice before declaring a structure unlawful?

❑ Does the inspector need to have the premises secured because the owner refuses or has failed to do so?

❑ Has the condemnation notice been posted in a conspicuous place on the structure or equipment?

❑ Has a photograph been taken of the posted notice?

❑ Has the condemnation notice been served on the responsible party prior to placarding the structure or equipment as "Condemned?"

❑ Has the property been placarded as "Condemned?"

❑ Has a photograph been taken of the "Condemned" placard?

❑ If the placard has been removed from the condemned property, does the inspector have enough evidence to prosecute someone for the violation? (e.g., eyewitness testimony, an admission by a defendant)

❑ If the occupant refuses to vacate the structure, should he or she be prosecuted?

❑ Is a referral to a social service agency needed in this case?

ORDER FOR CLOSING OF VACANT STRUCTURE—IPMC 108.2

Department of Property Maintenance

ORDER OF CONDEMNATION

THIS VACANT STRUCTURE IS UNFIT FOR HUMAN HABITATION AND OCCUPANCY AND ITS OCCUPANCY IS PROHIBITED BY THE CODE OFFICIAL. IT SHALL BE UNLAWFUL FOR ANY PERSON TO ENTER SUCH STRUCTURE WITHOUT THE APPROVAL OF THE CODE OFFICIAL. IT IS HEREBY ORDERED THAT THE STRUCTURE SHALL BE CLOSED UP SO AS NOT TO BE AN ATTRACTIVE NUISANCE.

125 S. 55th Street, WOODRIDGE, ILLINOIS
_____
PROPERTY ADDRESS

THIS NOTICE HAS BEEN POSTED ON THIS THE 30th DAY OF August 2006

BY: *Joan Rogers.*

ANY PERSON WHO DEFACES OR REMOVES A CONDEMNATION PLACARD WITHOUT THE APPROVAL OF THE CODE OFFICIAL SHALL BE SUBJECT TO *A FINE OF $750.00.**

John Black   *August 30, 2006*

CODE OFFICIAL   DATE

*Insert the penalty for occupying the premises, operating the equipment, or removing the placard.

NOTICE OF CONDEMNATION OF UNSAFE STRUCTURE—IPMC 108.3

NOTICE OF CONDEMNATION

Date: *June 28, 2006*

To: *Mr. Joseph McKenna**
 1020 Peal Rd.
 Woodridge, Illinois 60517

Re: *1020 Peal Rd.*
 Woodridge, Illinois

Based on the inspection of your property, *1020 Peal Rd., Woodridge*, Illinois, on *June 28, 2006*, I hereby condemn the structure because it is an unsafe structure pursuant to 2006 IPMC 108.1.1 in that it is dangerous to the life, health, property, or safety of the public or the occupants of the structure, because it fails to provide minimum safeguards to protect or warn occupants in the event of fire.** Therefore, I am serving you with this notice of condemnation.

As property owner you are hereby being directed to correct violations as noted below no later than the close of business on *July 14, 2006*.

FAILURE TO PROVIDE MEANS OF EGRESS: 2006 IPMC Section 702.1 as amended and adopted by reference in Section *8-1J-1(A)* of the *Village of Woodridge* Code—*You are required to provide a safe, continuous, and unobstructed path of travel from any point in the building or structure to the public way. The rooms are currently filled with newspapers, boxes, and other paper materials that block egress in the rooms to the outer doors.*

The property will be inspected on *July 16, 2006*. If it has been determined that the violations have not been corrected, the property shall be placarded as CONDEMNED and citations will be issued requiring your appearance in *DuPage County* Court.

You have a right to appeal this notice and order by filing a written application for appeal with the Board of Appeals for the *Village of Woodridge*. The application for appeal must be filed within twenty (20) days after the day this notice is served upon you. The appeal shall be based on a claim that the true intent of the code or the rules legally adopted thereunder have been incorrectly interpreted, the provisions of the code do not fully apply, or the requirements of the code are adequately satisfied by other means.

(Continued)

NOTICE OF CONDEMNATION OF UNSAFE STRUCTURE—IPMC 108.3 (*CONTINUED*)

If you fail to correct these violations, any action taken by the *Village of Woodridge*, the authority having jurisdiction, may be charged against the real estate upon which the structure is located and shall be a lien upon such real estate.

Very truly yours

Karyn Byrne

Deputy Code Official

Cc: Property File

*In the event of multiple responsible parties, all parties should be given a notice e.g., spouse.

**Or, because the structure contains unsafe equipment; or is so damaged, decayed, dilapidated, structurally unsafe, or of such faulty construction or unstable foundation that partial or complete collapse is possible.

COMMENT: This notice must also be posted in a conspicuous place in or about the structure and must be served on the owner or persons responsible for the property. The inspector should take a photograph of the posted notice so it cannot be argued later that it was never done.

SAMPLE FORM

NOTICE OF CONDEMNATION OF UNSAFE EQUIPMENT—IPMC 108.3

NOTICE OF CONDEMNATION

Date: *June 28, 2006*

To: *Mr. Joseph McKenna**
      1020 Peal Rd.
      Woodridge, Illinois 60517

Re: *1020 Peal Rd.*
      Woodridge, Illinois

Based on the inspection of your property, *1020 Peal Rd., Woodridge*, Illinois on *June 28, 2006*, I hereby condemn equipment,** specifically the *heating* equipment located therein, pursuant to *2006 IPMC-108.1.2* as amended and adopted by reference in Section *8-1J-1(A)* of the *Village of Woodridge* Code because it is a hazard to life, health, property, or safety of the public or occupants of the premises or structure. Therefore, I am serving you with this notice of condemnation.

As property owner you are hereby being directed to correct the violation as noted below no later than the close of business on *July 14, 2006*.

FAILURE TO MAINTAIN PLUMBING FIXTURE: 2006 IPMC Section 504.1 as amended and adopted by reference in Section *8-1J-1(A)* of the *Village of Woodridge* Code—*You are required to maintain the plumbing fixture in a safe working condition. The hot water heater in the basement is leaking gas.*

The equipment will be inspected on *July 17, 2006*. If it has been determined that the violations have not been corrected the property shall be placarded as CONDEMNED and citations will be issued requiring your appearance in *DuPage County* Court.

You have a right to appeal this notice and order by filing a written application for appeal with the Board of Appeals for the *Village of Woodridge*. The application for appeal must be filed within twenty (20) days after the day this notice is served upon you. The appeal shall be based on a claim that the true intent of the code or the rules legally adopted thereunder have been incorrectly interpreted, the provisions of the code do not fully apply, or the requirements of the code are adequately satisfied by other means.

If you fail to correct these violations, any action taken by the *Village of Woodridge*, the authority having jurisdiction, may be charged against the real estate upon which the structure is located and shall be a lien upon such real estate.

Very truly yours

Karyn Byrne

Deputy Code Official

Cc: Property File

*In the event of multiple responsible parties, all parties should be given a notice, e.g., spouse.

**Or, boiler, elevator, moving stairway, electrical wiring or device, flammable liquid containers, or other equipment on the premises or within the structure that is in such disrepair or condition that such equipment is a hazard to life, health, property, or safety of the public or occupants of the premises or structure.

NOTICE OF CONDEMNATION OF UNFIT FOR HUMAN OCCUPANCY—2006 IPMC 108.3

NOTICE OF CONDEMNATION

Date: *June 28, 2006*

To: *Mr. Joseph McKenna**
       1020 Peal Rd.
       Woodridge, Illinois 60517

Re: *1020 Peal Rd.*
      Woodridge, Illinois

Based on the inspection of your property, *1020 Peal Rd., Woodridge, Illinois* on *June 28, 2006,* I hereby condemn the structure because it is unfit for human occupancy pursuant to *2006 IPMC-108.1.3* because the structure is unsafe and unsanitary.** Therefore, I am serving you with this notice of condemnation.

As property owner you are hereby being directed to correct violations as noted below no later than the close of business on *August 14, 2006.*

1. FAILURE TO MAINTAIN INTERIOR STRUCTURE: 2006 IPMC Section 304.1 as amended and adopted by reference in Section *8-1J-1(A)* of the *Village of Woodridge* Code. *The interior structure must be maintained in good repair, structurally sound, and in sanitary condition. At present the ceiling in the living room is collapsing and mold is growing all over dishes and food in the kitchen.*

2. ACCUMULATION OF RUBBISH AND GARBAGE: 2006 IPMC Section 307.1 as amended and adopted by reference in Section *8-1J-1(A)* of the *Village of Woodridge* Code. *The exterior property and premises and the interior of every structure must be free from any accumulation of rubbish or garbage. Presently there are bags of garbage in the kitchen blocking the windows, torn bags of rubbish in the backyard, and piles of newspapers, boxes and other paper materials stacked to the ceiling in all of the rooms.*

3. EXTERMINATION: 2006 IPMC Section 308.1 as amended and adopted by reference in Section *8-1J-1(A)* of the *Village of Woodridge* Code: *The structure must be kept free from insect and rodent infestation. Evidence of cockroach infestation was found during the inspection of June 28, 2006.*

The property will be inspected on *August 17, 2006.* If it has been determined that the violations have not been corrected the property shall be placarded as CONDEMNED and citations will be issued requiring your appearance in *DuPage County* Court.

You have a right to appeal this notice and order by filing a written application for appeal with the Board of Appeals for the *Village of Woodridge.* The application for appeal must be filed within twenty (20) days after the day this notice is served upon you. The appeal shall be based on a claim that the true intent of the code or the rules legally adopted thereunder have been incorrectly interpreted, the provisions of the code do not fully apply, or the requirements of the code are adequately satisfied by other means.

(Continued)

If you fail to correct these violations, any action taken by the *Village of Woodridge*, the authority having jurisdiction, may be charged against the real estate upon which the structure is located and shall be a lien upon such real estate.

Very truly yours

Karyn Byrne

Deputy Code Official

Cc: Property File

*In the event of multiple responsible parties, all parties should be given a notice, e.g., spouse.
**Or, unlawful, or because of the degree to which the structure is in disrepair or lacks maintenance, vermin or rat infested, contains filth and contamination, or lacks ventilations, illumination, sanitary or heating facilities, or other essential equipment required by this code, or because the location of the structure constitutes a hazard to the occupants of the structure or to the public.

- -
NOTICE OF CONDEMNATION OF UNLAWFUL STRUCTURE—IPMC 108.3

NOTICE OF CONDEMNATION

Date: *June 28, 2006*

To: *Mr. Joseph McKenna**
1020 Peal Rd.
Woodridge, Illinois 60517

Re: *1020 Peal Rd.*
Woodridge, Illinois

Based on the inspection of your property, *1020 Peal Rd., Woodridge*, Illinois on *June 28, 2006*, I hereby condemn the structure located thereon, being the *garage*, pursuant to *2006 IPMC-108.1.4* as amended and adopted by reference in Section *8-1J-1(A)* of the *Village of Woodridge* Code because it was erected** contrary to law, in that *it was erected without a building permit and occupied without a certificate of occupancy*. Therefore, I am serving you with this notice of condemnation.

As property owner you are hereby being directed to correct the violation as noted below no later than the close of business on *July 26, 2006* by either removing the unlawful structure or obtaining the required building permits, having the necessary inspections, and obtaining a certificate of occupancy.

The structure will be inspected on *July 27, 2006*. If it has been determined that the violations have not been corrected, the structure shall be placarded as CONDEMNED and citations will be issued requiring your appearance in *DuPage County* Court.

You have a right to appeal this notice and order by filing a written application for appeal with the Board of Appeals for the *Village of Woodridge*. The application for appeal must be filed within twenty (20) days after the day this notice is served upon you. The appeal shall be based on a claim that the true intent of the code or the rules legally adopted thereunder have been incorrectly interpreted, the provisions of the code do not fully apply, or the requirements of the code are adequately satisfied by other means.

If you fail to correct these violations, any action taken by the *Village of Woodridge*, the authority having jurisdiction, may be charged against the real estate upon which the structure is located and shall be a lien upon such real estate.

Very truly yours

Karyn Byrne

Deputy Code Official

Cc: Property File

*In the event of multiple responsible parties, all parties should be given a notice, e.g., spouse.
**Or, altered or occupied.

SAMPLE FORM

CONDEMNATION PLACARD—IPMC 108.4

Department of Property Maintenance

NOTICE CONDEMNED

THIS STRUCTURE* IS UNSAFE** AND ITS OCCUPANCY IS PROHIBITED BY THE CODE OFFI-CIAL. IT SHALL BE UNLAWFUL FOR ANY PERSON TO OCCUPY SUCH STRUCTURE WITH-OUT THE APPROVAL OF THE CODE OFFICIAL

1020 PEAL ROAD, WOODRIDGE, ILLINOIS

_____

PROPERTY ADDRESS

THIS NOTICE HAS BEEN POSTED ON THIS THE *17th* DAY OF *JULY 2006*

BY: *Karyn Byrne.*

ANY PERSON WHO OCCUPIES THE PREMISES,* OR DEFACES OR REMOVES A CONDEMNATION PLACARD WITHOUT THE APPROVAL OF THE CODE OFFICIAL SHALL BE SUBJECT TO** *A FINE OF $750.00.*****

John Black   *July 17, 2006*

CODE OFFICIAL   DATE

  *Or, equipment.

 **Or, unfit for human occupancy or unlawful.

***Or, operates such equipment.

****Insert penalty for occupying the premises, operating the equipment, or removing the placard.

UNLAWFUL REMOVAL OF A PLACARD—IPMC 108.4.1

STATE OF *ILLINOIS*
COUNTY OF DUPAGE
VILLAGE OF WOODRIDGE
v.
NAME: *JOSEPH McKENNA*
ADDRESS: *1020 Peal Road*
CITY: *Woodridge, Illinois 60517*

The undersigned says that on or about *July 18, 2006*, at or about *3:00 p.m.* the Defendant, *Joseph McKenna*, did unlawfully commit the offense of **Unlawful Removal of a Condemnation Placard** in violation of *2006* **IPMC-108.4.1** as amended and adopted by reference in Section *8-1J-1(A)* of the *Village of Woodridge* Code; in that said Defendant, without the approval of the code official, *John Black*, did remove a condemnation placard that had been posted on the structure* at *1020 Peal Road, Woodridge, Illinois*, on *July 17, 2006*.

Karyn Byrne

Complainant

Sworn to and Subscribed before Me
This *18th* Day of *July, 2006*

_____

Notary Public

*Or, equipment.

COMMENT: Always take a photograph of the placard after it is posted as proof that the inspector properly posted the property.

OCCUPYING* A PLACARDED STRUCTURE—IPMC 108.5

STATE OF *ILLINOIS*
COUNTY OF DUPAGE
VILLAGE OF WOODRIDGE
v.
NAME: *JOSEPH McKENNA*
ADDRESS: *1020 Peal Road*
CITY: *Woodridge, Illinois 60517*

The undersigned says that on or about *July 18, 2006*, at or about *3:00 p.m.* the Defendant, *Joseph McKenna*, being the owner** of *1020 Peal Road*, did unlawfully commit the offense of **Occupying*** a Placarded Structure** in violation of *2006* **IPMC-108.5** as amended and adopted by reference in Section *8-1J-1(A)* of the *Village of Woodridge* Code; in that said Defendant, did occupy**** a placarded structure, being *1020 Peal Road, Woodridge, Illinois,* which had been placarded as "Condemned" on *July 17, 2006* by the code official, *John Black,* of the *Village of Woodridge.*

Karyn Byrne

Complainant

Sworn to and Subscribed before Me
This *18th* Day of *July, 2006*

_____

Notary Public

  *Or, Allowing the Occupation of
 **Or, person responsible for the premises
***Or, Allowing the Occupation of
****Or, allowed the occupation of

OPERATING* PLACARDED EQUIPMENT—IPMC 108.5

STATE OF *ILLINOIS*
COUNTY OF DUPAGE
VILLAGE OF WOODRIDGE
v.
NAME: *JOSEPH McKENNA*
ADDRESS: *1020 Peal Road*
CITY: *Woodridge, Illinois 60517*

The undersigned says that on or about *July 18, 2006*, at or about *3:00 p.m.* the Defendant, *Joseph McKenna*, being the owner** of 1020 Peal Road, did unlawfully commit the offense of **Operating* Placarded Equipment** in violation of *2006* **IPMC-108.5** as amended and adopted by reference in Section *8-1J-1(A)* of the *Village of Woodridge* Code, in that said Defendant, did operate*** placarded equipment, being *a hot water heater at 1020 Peal Road, Woodridge, Illinois*, which had been placarded as "Condemned" on *July 17, 2006* by the code official, *John Black*, of the *Village of Woodridge*.

Karyn Byrne

Complainant

Sworn to and Subscribed before Me
This *18th* Day of *July, 2006*

_____

Notary Public

*Or, Allowing the Operation of
**Or, person responsible for the premises
***Or, allowed the operation of

IPMC 109   Emergency Measures

NOTICE OF IMMINENT DANGER—IPMC 109.1

Department of Property Maintenance

NOTICE AND ORDER

THIS STRUCTURE IS UNSAFE AND ITS OCCUPANCY IS PROHIBITED BY THE CODE OFFICIAL. ALL OCCUPANTS MUST VACATE THE PREMISES FORTHWITH.

IT SHALL BE UNLAWFUL FOR ANY PERSON TO ENTER SUCH STRUCTURE EXCEPT FOR THE PURPOSE OF SECURING THE STRUCTURE, MAKING THE REQUIRED REPAIRS, REMOVING THE HAZARDOUS CONDITION, OR DEMOLISHING THE SAME BUT ONLY AFTER HAVING SECURED THE REQUIRED PERMITS.

1020 PEAL ROAD, WOODRIDGE, ILLINOIS

PROPERTY ADDRESS

THIS NOTICE HAS BEEN POSTED ON THIS THE *17th* DAY OF *JULY 2006*

John Black   *July 17, 2006*

CODE OFFICIAL   DATE

You have a right to a hearing to appeal this notice and order by filing a written application for appeal with the Board of Appeals for the *Village of Woodridge*. The application for appeal must be filed within twenty (20) days after the day this notice is served upon you. The appeal shall be based on a claim that the true intent of the code or the rules legally adopted thereunder have been incorrectly interpreted, the provisions of the code do not fully apply, or the requirements of the code are adequately satisfied by other means.

IPMC 110   Demolition

CHECKLIST

DEMOLITION

❑ Does the inspector possess the necessary expertise to find that the structure needs to be demolished?

❑ Should the inspector refer the matter to the building code official or obtain the services of an expert before finding that the structure needs to be demolished?

❑ Does the inspector need to have the premises demolished because the owner refuses or has failed to do so?

❑ Has the notice of demolition been served properly on the responsible party?

❑ Should a referral to a social service agency be made in this case?

❑ Has legal counsel been consulted regarding the necessary legal steps that must be taken in a demolition case?

❑ Has the case been referred to legal counsel?

❑ Has a court action for demolition been filed?

❑ Will a lien be placed on the property after demolition?

❑ Who will be responsible for filing the lien?

SAMPLE FORM

NOTICE OF DEMOLITION (NO REPAIR POSSIBLE)—IPMC 110.1

September 5, 2006

Mr. Joseph McKenna
1020 Peal Road
Woodridge, IL 60517

Re: 1020 Peal Road, Woodridge, Illinois

Dear *Mr. McKenna*:

An inspection of your property at *1020 Peal Road. Woodridge, Illinois*, on *September 1, 2006* shows that *an accessory* structure, being *the garage,* is so old, dilapidated, or has become so out of repair as to be dangerous,* to the extent that it is unreasonable to repair the structure in that *the roof and part of a wall have collapsed.* Pursuant to *2006* IPMC 110.1 as amended and adopted by reference in Section *8-1J-1(A)* of the *Village of Woodridge* Code, it is hereby ordered that the structure must be demolished and removed by *October 3, 2006.*

You have a right to appeal this notice and order by filing a written application for appeal with the Board of Appeals for the *Village of Woodridge.* The application for appeal must be filed within twenty (20) days after the day this notice is served upon you. The appeal shall be based on a claim that the true intent of the code or the rules legally adopted thereunder have been incorrectly interpreted, the provisions of the code do not fully apply, or the requirements of the code are adequately satisfied by other means.

If you fail to demolish and remove the structure, the *Village of Woodridge,* through the code official, *John Black,* will cause the structure to be demolished and removed, either through an available public agency or by contract or arrangement with private persons, and the costs of such demolition and removal shall be charged against the real estate upon which the structure is located and shall be a lien upon the real estate.

Please feel free to contact me to discuss this matter further.

Very truly yours

Karyn Byrne

Deputy Code Official

*Or, unsafe, unsanitary or otherwise unfit for human habitation or occupancy

----

NOTICE OF DEMOLITION (REPAIR POSSIBLE)—IPMC 110.1

September 5, 2006

Mr. Joseph McKenna
1020 Peal Road
Woodridge, IL 60517

Re: 1020 Peal Road, Woodridge, Illinois

Dear *Mr. McKenna*:

An inspection of your property at *1020 Peal Road. Woodridge, Illinois*, on *September 1, 2006* shows that the structure, being *the residence,* is unsafe* *due to extensive damage to the walls by a fire.* Pursuant to *2006* IPMC 110.1 as amended and adopted by reference in Section *8-1J-1(A)* of the *Village of Woodridge* Code, it is hereby ordered that the structure must be repaired and made safe and sanitary, or demolished and removed at your option by *October 3, 2006*. All required permits must be obtained before any repair or demolition begins.

You have a right to appeal this notice and order by filing a written application for appeal with the Board of Appeals for the *Village of Woodridge*. The application for appeal must be filed within twenty (20) days after the day this notice is served upon you. The appeal shall be based on a claim that the true intent of the code or the rules legally adopted thereunder have been incorrectly interpreted, the provisions of the code do not fully apply, or the requirements of the code are adequately satisfied by other means.

If you fail to repair and make safe or demolish and remove the structure, the *Village of Woodridge*, through the code official, *John Black*, will cause the structure to be demolished and removed, either through an available public agency or by contract or arrangement with private persons, and the costs of such demolition and removal shall be charged against the real estate upon which the structure is located and shall be a lien upon the real estate.

Please feel free to contact me to discuss this matter further.

Very truly yours

Karyn Byrne

Deputy Code Official

*Or, dangerous, unsanitary or otherwise unfit for human habitation or occupancy.

NOTICE OF DEMOLITION (CESSATION OF NORMAL CONSTRUCTION)—IPMC 110.1

September 5, 2006

Mr. Mark Sucher
738 Meadow Lane
Woodridge, IL 60517

Re: 738 *Meadow Lane, Woodridge, Illinois*

Dear *Mr. Sucher*:

An inspection of your property at *738 Meadow Lane. Woodridge, Illinois,* on *September 2, 2006,* shows that the structure, being a *room addition under construction,* is so out of repair as to be dangerous* *due to your failure to enclose the structure after framing.* There has been a cessation of normal construction on the structure since *July 8, 2004.* Pursuant to *2006* IPMC 110.1 as amended and adopted by reference in Section *8-1J-1(A)* of the *Village of Woodridge* Code, it is hereby ordered that the structure must be demolished and removed by *October 1, 2006.* All required permits must be obtained before any demolition begins.

You have a right to appeal this notice and order by filing a written application for appeal with the Board of Appeals for the *Village of Woodridge.* The application for appeal must be filed within twenty (20) days after the day this notice is served upon you. The appeal shall be based on a claim that the true intent of the code or the rules legally adopted thereunder have been incorrectly interpreted, the provisions of the code do not fully apply, or the requirements of the code are adequately satisfied by other means.

If you fail to demolish and remove the structure, the *Village of Woodridge,* through the code official, *John Black,* will cause the structure to be demolished and removed, either through an available public agency or by contract or arrangement with private persons, and the costs of such demolition and removal shall be charged against the real estate upon which the structure is located and shall be a lien upon the real estate.

Please feel free to contact me to discuss this matter further.

Very truly yours

Karyn Byrne

Deputy Code Official

*Or, unsafe, unsanitary or otherwise unfit for human habitation or occupancy.

CHAPTER 2

Definitions

Legal definitions are meant to be very precise. Therefore, it is important to read the definitions in the IPMC carefully to see if they apply to the inspector's case. The following sections are designed to help the inspector analyze the facts to see if a particular definition applies to the situation being investigated.

Definitions

I. WHO IS THE OWNER?

A. Any of the following:
   - ❑ Person ❑ Firm
   - ❑ Agent ❑ Corporation

B. And (choose one):
   - ❑ Has a legal interest in the property
   - ❑ Has an equitable interest in the property
   - ❑ Has title based on a document recorded in the official records of the state, county, or municipality
   - ❑ Has control of the property (e.g., guardian, executor, administrator of an estate if ordered to take possession of real property by a court)

COMMENT:

A. One of the biggest mistakes inspectors make is not realizing that an owner can be a legal entity, such as a corporation, instead of a human being.

B. There must be a direct connection between the person or entity and the property.

II. IS THE INDIVIDUAL AN OCCUPANT?

An occupant:

(Choose any of the following)
   - ❑ Lives in a building
   - ❑ Sleeps in a building
   - ❑ Has possession of a space within a building

COMMENT: Occupants can often be responsible for the interior of the property, so it is important that they be properly identified.

III. WHO IS A TENANT?

Any of the following:
   - ❑ Person
   - ❑ Corporation
   - ❑ Partnership   +   Occupies a building
   - ❑ Group   or portion thereof

COMMENT: A tenant does not have to be a legal owner of record. A tenant is not necessarily a human being and can be a legal entity, such as a corporation.

IV. WHO IS A PERSON UNDER THE CODE?

Any of the following:
   - ❑ Individual
   - ❑ Corporation
   - ❑ Partnership
   - ❑ Group acting as a unit

COMMENT: A "person" is not always an individual.

V. WHO IS AN OPERATOR?

A person who has:

❑ Charge, care, or control of a structure or premises

❑ The structure or premises is let or offered for occupancy

COMMENT: The term "operator" is often used to describe a management agent employed by an owner.

VI. IS IT GARBAGE OR RUBBISH?

Rubbish

❑ Combustible waste material, not garbage

❑ Noncombustible waste material, not garbage

For example:

❑ Residue from burning of wood, coal, coke, and other combustible material	❑ Cartons	❑ Wood
	❑ Rubber	❑ Glass
	❑ Excelsior	❑ Boxes
	❑ Crockery	❑ Other similar material
❑ Paper	❑ Mineral matter	❑ Tin cans
❑ Leather	❑ Dust	❑ Tree branches
	❑ Rags	❑ Yard trimmings

Garbage

❑ Animal waste or

❑ Vegetable waste resulting from the handling, preparation, and consumption of food

COMMENT: Never substitute "junk" or "debris" for the more specific term "rubbish" or "garbage" as each term has a very specific meaning.

VII. Table 2-1 WHAT KIND OF LIVING UNIT IS IT?

	Single Unit	Room or Group of Rooms Forming Single Habitable Space	One or More Persons	Complete and Independent	Eating and Cooking Facilities	Sanitary Facilities	Sleeping and Living Area
Dwelling Unit	Yes	*	Yes	Yes	Yes	Yes	Yes
Housekeeping Unit	*	Yes	*	No	Yes	No	Yes
Rooming Unit	*	Yes	*	No	No**	*	Yes
Sleeping Unit*	*	*	Yes	No	Eating— Yes Cooking— yes, if no sanitation facilities	Yes, if no kitchen facilities	Yes

*Indicates that the code definition is silent as to this element.
**2006 IPMC allows cooking devices such as coffee pots and microwave ovens.
***Rooms and spaces that are also part of a dwelling unit are not sleeping rooms.

COMMENT: Given the variety of living units, it is important to know how they differ.

VIII. WHO IS THE PARTY RESPONSIBLE FOR THE VIOLATIONS?

Determining who the responsible party is for a violation may be the single most important task for the inspector, second only to the violation itself. Sometimes, more than one party must be notified and taken to court. This table summarizes the responsibility sections at the beginning of every chapter of the IPMC and compares those provisions side by side. The terms "owner," "occupant," and "person" are defined in Chapter 2. They carry very specific meanings. "Owner should be occupant" is not defined but clearly means an owner who lives or sleeps in a building or has possession of a space within a building. "Person" can include an individual, corporation, partnership, or any other group acting as a unit; as well as an owner or occupant, for example, a building manager, a management company or an agent. If there is a lease, the inspector should attempt to see a copy of it to determine if there are specific references to the duties or responsibilities of the landlord or tenant, and to determine who has control over various portions of the premises (e.g., accessory structure, such as a garage).

Table 2-2 WHO IS THE PARTY RESPONSIBLE FOR THE VIOLATION?

	Owner	Owner-Occupant	Person	Occupant
Structure and exterior property violations—IPMC 301.2	X			
Occupying an unsanitary and unsafe premises—IPMC 301.2		X		

(Continued)

Table 2-2 WHO IS THE PARTY RESPONSIBLE FOR THE VIOLATION? (*CONTINUED*)

	Owner	Owner-Occupant	Person	Occupant
Allowing another to occupy an unsanitary and unsafe premises—IPMC 301.2			X	
Premises which are controlled by an occupant and occupied in an unsanitary and unsafe condition—IPMC 301.2				X
Light, ventilation, or occupancy limitation violations—IPMC 401.2	X			
Occupying a premises with light, ventilation, or occupancy limitation violations—IPMC 401.2		X		
Permitting occupancy of a premises with light, ventilation or occupancy limitation violations—IPMC 401.2			X	
Plumbing violations—IPMC 501.2	X			
Occupying a premises with plumbing violations—IPMC 601.2		X		
Permitting the occupancy of a premises with plumbing violations—IPMC 601.2			X	
Mechanical and electrical violations—IPMC 601.2	X			
Occupying a premises with mechanical and electrical violations—IPMC 601.2		X		
Permitting the occupancy of a premises with mechanical and electrical violations—IPMC 601.2			X	
Fire safety violations—IPMC 701.2	X			
Occupying a premises with fire safety violations—IPMC 701.2		X		
Permitting the occupancy of a premises with fire safety violations—IPMC 701.2			X	

Maintenance of Exterior and Interior Property Areas

CHAPTER 3 covers the minimum conditions for the maintenance of structures, equipment, and exterior property, and the responsibilities that persons have for that maintenance. The chapter prescribes minimum standards for the maintenance of exterior property, swimming pools, spas, hot tubs, exterior structures, interior structures, and handrails and guardrails. It regulates rubbish, garbage, and the extermination of insects and rodents. Because of its comprehensive nature, it is the most referenced of all of the chapters in the International Property Maintenance Code.

IPMC 301   General Violations

SAMPLE COMPLAINT

OCCUPYING UNSANITARY AND UNSAFE PREMISES—IPMC 301.2

STATE OF *ILLINOIS*
COUNTY OF DUPAGE
VILLAGE OF WOODRIDGE
v.
NAME: *ROBERT MEYER*
ADDRESS: *4617 Carousel St.*
CITY: *Woodridge, Illinois 60517*

The undersigned says that on or about *April 2, 2006*, at or about *3:00 p.m.* the Defendant, *Robert Meyer*, did unlawfully commit the offense **Occupying Unsanitary and Unsafe Premises** in violation of *2006* **IPMC-301.2** as amended and adopted by reference in Section *8-1J-1(A) of the Village of Woodridge* Code; in that said Defendant, the owner-occupant, occupied the premises of *4617 Carousel St., Woodridge, Illinois*, at a time when the premises were in violation of *2006 IPMC-304.7* as amended and adopted by reference in Section *8-1J-1(A) of the Village of Woodridge* Code in that *the roof has not been maintained in a sound, and tight condition and has defects, being missing shingles, that admit rain into the interior of the structure, which has caused mold to grow in the interior.*

Joan Rogers

Complainant

Sworn to and Subscribed before Me
This *30th* Day of *April, 2006*

_____

Notary Public

ALLOWING PERSONS TO OCCUPY AN UNSANITARY AND UNSAFE PREMISES—IPMC 301.2

STATE OF *ILLINOIS*
COUNTY OF DUPAGE
VILLAGE OF WOODRIDGE
v.
NAME: *PLEASANT DREAMS MANAGEMENT CO.*
ADDRESS: *34 W. Reber St.*
CITY: *Woodridge, Illinois 60000*

The undersigned says that on or about *November 18, 2006*, at or about *3:00 p.m.* the Defendant, *Pleasant Dreams Management Co., the managing agent for the property*, did unlawfully commit the offense **Allowing Persons to Occupy an Unsanitary and Unsafe Premises** in violation of *2006* **IPMC-301.2** as amended and adopted by reference in Section *8-1J-1(A)* of the *Village of Woodridge* Code; in that said Defendant did permit *Joan Knueven* to occupy the premises of *34 W. Reber St., Unit 7 G, Woodridge, Illinois,* at a time when the premises were in violation of *2006 IPMC-304.7* as amended and adopted by reference in Section *8-1J-1(A)* of the *Village of Woodridge* Code in that *the roof has not been maintained in a sound and tight condition, and has defects, being missing shingles, that admit rain into the interior of the structure, which has caused mold to grow on the ceiling of the apartment.*

Joan Rogers
Complainant

Sworn to and Subscribed before Me
This *8th* Day of *December, 2006*

_____
Notary Public

SAMPLE COMPLAINT

FAILURE TO MAINTAIN VACANT STRUCTURE—IPMC 301.3

STATE OF *ILLINOIS*
COUNTY OF DUPAGE
VILLAGE OF WOODRIDGE
v.
NAME: *ROBERT MEYER*
ADDRESS: *4617 Carousel St.*
CITY: *Woodridge, Illinois 60517*

The undersigned says that on or about *April 16, 2006*, at or about *3:00 p.m.* the Defendant did unlawfully commit the offense of **Failure to Maintain Vacant Structure** in violation of *2006* **IPMC-301.3** as amended and adopted by reference in Section *8-1J-1(A)* of the *Village of Woodridge* Code, in that said Defendant, the owner of *4617 Carousel St., Woodridge, Illinois*, failed to maintain a vacant structure* at *4617 Carousel St., Woodridge, Illinois*, in a clean, safe, secure, and sanitary condition so as not to cause a blighting problem or adversely affect the public health or safety in that *the residence has broken windows, the front door is not secure, and the basement windows have been removed.*

Joan Rogers
Complainant

Sworn to and Subscribed before Me
This *30th* Day of *April, 2006*

_____
Notary Public

*Or, the premises.

SAMPLE COMPLAINT

FAILURE TO MAINTAIN VACANT LAND—IPMC 301.3

STATE OF *ILLINOIS*
COUNTY OF DUPAGE
VILLAGE OF WOODRIDGE
v.
NAME: *ROBERT MEYER*
ADDRESS: *4617 Carousel St.*
CITY: *Woodridge, Illinois 60517*

The undersigned says that on or about *April 16, 2006*, at or about *3:00 p.m.* the Defendant did unlawfully commit the offense of **Failure to Maintain Vacant Land** in violation of *2006* **IPMC-301.3** as amended and adopted by reference in Section *8-1J-1(A)* of the *Village of Woodridge* Code, in that said Defendant, the owner of *4619 Carousel St., Woodridge, Illinois*, failed to maintain vacant land, in a clean, safe, secure, and sanitary condition so as not to cause a blighting problem or adversely affect the public health or safety in that *the lot has weeds in excess of 8 inches and there are torn bags of rubbish and garbage strewn about the property.*

Joan Rogers
Complainant

Sworn to and Subscribed before Me
This *30th* Day of *April, 2006*

_____

Notary Public

IPMC 302 Exterior Property Area Violations

SAMPLE COMPLAINT

FAILURE TO MAINTAIN EXTERIOR PROPERTY—IPMC 302.1

STATE OF *ILLINOIS*
COUNTY OF DUPAGE
VILLAGE OF WOODRIDGE
v.
NAME: *ROBERT MEYER*
ADDRESS: *4617 Carousel St.*
CITY: *Woodridge, Illinois 60517*

The undersigned says that on or about *April 2, 2006*, at or about *3:00 p.m.* the Defendant did unlawfully commit the offense of **Failure to Maintain Exterior Property** in violation of **2006 IPMC-302.1** as amended and adopted by reference in Section *8-1J-1(A)* of the *Village of Woodridge* Code, in that said Defendant, the owner of *4617 Carousel St., Woodridge, Illinois*, failed to maintain the exterior of the property in a clean, safe, and sanitary condition in that *there were piles of garbage, rusted machine parts, and dead tree limbs in the backyard.*

Joan Rogers
Complainant

Sworn to and Subscribed before Me
This *30th* Day of *April, 2006*

_____

Notary Public

FAILURE TO MAINTAIN GRADING AND DRAINAGE—IPMC 302.2

STATE OF *ILLINOIS*

COUNTY OF DUPAGE

VILLAGE OF WOODRIDGE

v.

NAME: *ROBERT MEYER*

ADDRESS: *4617 Carousel St.*

CITY: *Woodridge, Illinois 60517*

The undersigned says that on or about *April 2, 2006*, at or about *3:00 p.m.* the Defendant did unlawfully commit the offense of **Failure to Maintain Grading and Drainage** in violation of **2006 IPMC-302.2** as amended and adopted by reference in Section *8-1J-1(A)* of the *Village of Woodridge* Code, in that said Defendant, the owner of *4617 Carousel St., Woodridge, Illinois*, failed to maintain the grading and drainage on the premises to prevent the erosion of soil and to prevent the accumulation of stagnant water in that *there is no ground cover to prevent soil erosion and stagnant water is accumulating because of piles of dirt in the backyard.*

Joan Rogers

Complainant

Sworn to and Subscribed before Me

This *30th* Day of *April, 2006*

_____

Notary Public

FAILURE TO MAINTAIN SIDEWALKS AND DRIVEWAYS—IPMC 302.3

STATE OF *ILLINOIS*
COUNTY OF DUPAGE
VILLAGE OF WOODRIDGE
v.
NAME: *ROBERT MEYER*
ADDRESS: *4617 Carousel St.*
CITY: *Woodridge, Illinois 60517*

The undersigned says that on or about *April 2, 2006*, at or about *3:00 p.m.* the Defendant did unlawfully commit the offense of **Failure to Maintain Sidewalk*** in violation of ***2006 IPMC-302.3*** as amended and adopted by reference in Section *8-1J-1(A)* of the *Village of Woodridge* Code, in that said Defendant, the owner of *4617 Carousel St., Woodridge, Illinois*, failed to maintain the sidewalk* on the premises in a proper state of repair and free from hazardous conditions in that *the deteriorated concrete sidewalk is a trip hazard.*

Joan Rogers
Complainant

Sworn to and Subscribed before Me
This *30th* Day of *April, 2006*

_____
Notary Public

*Or, walkway, stairs, driveway, parking spaces.

COMMENT: When enforcing the section for a sidewalk, make sure it is on private property, not public property.

FAILURE TO MAINTAIN PROPERTY FREE FROM WEEDS—IPMC 302.4

STATE OF *ILLINOIS*
COUNTY OF DUPAGE
VILLAGE OF WOODRIDGE
v.
NAME: *ROBERT MEYER*
ADDRESS: *4617 Carousel St.*
CITY: *Woodridge, Illinois 60517*

The undersigned says that on or about *April 2, 2006*, at or about *3:00 p.m.* the Defendant did unlawfully commit the offense of **Failure to Maintain Property Free from Weeds** in violation of **2006 IPMC-302.4** as amended and adopted by reference in Section *8-1J-1(A)* of the *Village of Woodridge* Code, in that said Defendant, the owner of *4617 Carousel St., Woodridge, Illinois*, failed to maintain the premises and exterior property free from weeds* in excess of *8*** inches.

Joan Rogers
Complainant

Sworn to and Subscribed before Me
This *30th* Day of *April, 2006*

_____

Notary Public

 *Or, plant growth.
**Insert height of inches adopted by reference by the jurisdiction.

FAILURE TO MAINTAIN PROPERTY FREE FROM NOXIOUS WEEDS—IPMC 302.4

STATE OF *ILLINOIS*
COUNTY OF DUPAGE
VILLAGE OF WOODRIDGE
v.
NAME: *ROBERT MEYER*
ADDRESS: *4617 Carousel St.*
CITY: *Woodridge, Illinois 60517*

The undersigned says that on or about *April 2, 2006*, at or about *3:00 p.m.* the Defendant did unlawfully commit the offense of **Failure to Maintain Property Free from Noxious Weeds** in violation of ***2006* IPMC-302.4** as amended and adopted by reference in Section *8-1J-1(A)* of the *Village of Woodridge* Code, in that said Defendant, the owner of *4617 Carousel St., Woodridge, Illinois,* failed to maintain the premises and exterior property free from noxious weeds* *being thistles.*

Joan Rogers
Complainant

Sworn to and Subscribed before Me
This *30th* Day of *April, 2006*

_____
Notary Public

*Or, plant growth.

FAILURE TO PREVENT RODENT HARBORAGE—IPMC 302.5

STATE OF *ILLINOIS*
COUNTY OF DUPAGE
VILLAGE OF WOODRIDGE
v.
NAME: *ROBERT MEYER*
ADDRESS: *4617 Carousel St.*
CITY: *Woodridge, Illinois 60517*

The undersigned says that on or about *April 2, 2006,* at or about *3:00 p.m.* the Defendant did unlawfully commit the offense of **Failure to Prevent Rodent Harborage** in violation of **2006 IPMC-302.5** as amended and adopted by reference in Section *8-1J-1(A)* of the *Village of Woodridge* Code; in that said Defendant, the owner of *4617 Carousel St., Woodridge, Illinois*, failed to keep the exterior of the property* free from rodent infestation in that *rat droppings were present in the backyard by open bags of rotting garbage.*

Joan Rogers
Complainant

Sworn to and Subscribed before Me
This *30th* Day of *April, 2006*

_____
Notary Public

*Or, structure.

SAMPLE COMPLAINT

FAILURE TO EXTERMINATE RODENTS—IPMC 302.5

STATE OF *ILLINOIS*
COUNTY OF DUPAGE
VILLAGE OF WOODRIDGE
v.
NAME: *ROBERT MEYER*
ADDRESS: *4617 Carousel St.*
CITY: *Woodridge, Illinois 60517*

The undersigned says that on or about *April 2, 2006*, at or about *3:00 p.m.* the Defendant did unlawfully commit the offense of **Failure to Exterminate Rodents** in violation of ***2006 IPMC-302.5*** as amended and adopted by reference in Section *8-1J-1(A)* of the *Village of Woodridge* Code; in that said Defendant, the owner of *4617 Carousel St., Woodridge, Illinois*, failed to exterminate rodents after evidence of rodent harborage, being *rat droppings*, was found on the exterior of the property*.

Joan Rogers

Complainant

Sworn to and Subscribed before Me
This *30th* Day of *April, 2006*

_____

Notary Public

*Or, structure.

- -

FAILURE TO PREVENT REINFESTATION OF RODENTS—IPMC 302.5

STATE OF *ILLINOIS*
COUNTY OF DUPAGE
VILLAGE OF WOODRIDGE
v.
NAME: *ROBERT MEYER*
ADDRESS: *4617 Carousel St.*
CITY: *Woodridge, Illinois 60517*

The undersigned says that on or about *April 2, 2006*, at or about *3:00 p.m.* the Defendant did unlawfully commit the offense of **Failure to Prevent Reinfestation of Rodents** in violation of *2006* **IPMC-302.5** as amended and adopted by reference in Section *8-1J-1(A)* of the *Village of Woodridge* Code; in that said Defendant, the owner of *4617 Carousel St., Woodridge, Illinois,* failed to take proper precautions to eliminate rodent harborage and prevent reinfestation after extermination of the exterior of the property* in *that the backyard is filled with yard waste and bags of rotting food.*

Joan Rogers
Complainant

Sworn to and Subscribed before Me
This *30th* Day of *April, 2006*

_____
Notary Public

*Or, structure.

FAILURE TO MAINTAIN EXHAUST VENTS—IPMC 302.6

STATE OF *ILLINOIS*
COUNTY OF DUPAGE
VILLAGE OF WOODRIDGE
V.
NAME: *KYLE'S CHICKEN SHACK, INC.*
ADDRESS: *109 Fremont St.*
CITY: *Woodridge, Illinois 60517*

The undersignd says that on or about *April 2, 2006*, at or about *3:00 p.m.* the Defendant did unlawfully commit the offense of **Failure to Maintain Exhaust Vents** in violation of **2006 IPMC-302.6** as amended and adopted by reference in Section *8-1J-1(A)* of the *Village of Woodridge* Code, in that said Defendant, the owner of *109 Fremont St., Woodridge, Illinois*, failed to maintain the exhaust vents in that smoke* from a fan** *in the restaurant* located on the premises is being discharged directly upon adjacent public*** property, *being the Woodridge Village Park at 105 Fremont St., Woodridge, IL.*

Joan Rogers
Complainant

Sworn to and Subscribed before Me
This *30th* Day of *April, 2006*

_____
Notary Public

   *Or, gas, steam, vapor, hot air, grease, odors, or other gaseous or particulate wastes.
  **Or, pipe, duct, conductor, or blower.
***Or, private or that of another tenant.

FAILURE TO MAINTAIN ACCESSORY STRUCTURE—IPMC 302.7

STATE OF *ILLINOIS*
COUNTY OF DUPAGE
VILLAGE OF WOODRIDGE
v.
NAME: *ROBERT MEYER*
ADDRESS: *4617 Carousel St.*
CITY: *Woodridge, Illinois 60517*

The undersigned says that on or about *April 2, 2006,* at or about *3:00 p.m.* the Defendant did unlawfully commit the offense of **Failure to Maintain Accessory Structure** in violation of *2006* **IPMC-302.7** as amended and adopted by reference in Section *8-1J-1(A)* of the *Village of Woodridge* Code; in that said Defendant, the owner of *4617 Carousel St., Woodridge, Illinois,* failed to maintain the garage* upon such premises in good repair in that *the garage has holes in the roof and parts of the siding are missing.*

Joan Rogers
Complainant

Sworn to and Subscribed before Me
This *30th* Day of *April, 2006*

_____
Notary Public

*Or, fence, wall, or other accessory structure.

SAMPLE COMPLAINT

KEEPING AN INOPERATIVE MOTOR VEHICLE—IPMC 302.8

STATE OF *ILLINOIS*
COUNTY OF DUPAGE
VILLAGE OF WOODRIDGE
v.
NAME: *ROBERT MEYER*
ADDRESS: *4617 Carousel St.*
CITY: *Woodridge, Illinois 60517*

The undersigned says that on or about *April 2, 2006*, at or about *3:00 p.m.* the Defendant did unlawfully commit the offense of **Keeping an Inoperative Motor Vehicle** in violation of **2006 IPMC-302.8** as amended and adopted by reference in Section *8-1J-1(A)* of the *Village of Woodridge* Code; in that said Defendant, the owner of *4617 Carousel St., Woodridge, Illinois*, did maintain an inoperative motor vehicle on the premises in that *a blue sedan in the driveway has no engine and two flat tires.*

Joan Rogers
Complainant

Sworn to and Subscribed before Me
This *30th* Day of *April, 2006*

_____

Notary Public

KEEPING AN UNLICENSED MOTOR VEHICLE—IPMC 302.8

STATE OF *ILLINOIS*
COUNTY OF DUPAGE
VILLAGE OF WOODRIDGE
v.
NAME: *ROBERT MEYER*
ADDRESS: *4617 Carousel St.*
CITY: *Woodridge, Illinois 60517*

The undersigned says that on or about *April 2, 2006*, at or about *3:00 p.m.* the Defendant did unlawfully commit the offense of **Keeping an Unlicensed Motor Vehicle** in violation of *2006* **IPMC-302.8** as amended and adopted by reference in Section *8-1J-1(A)* of the *Village of Woodridge* Code; in that said Defendant, the owner of *4617 Carousel St., Woodridge, Illinois,* did maintain an unlicensed motor vehicle on the premises in that *a blue sedan in the driveway has expired registration plates from 2004.*

Joan Rogers
Complainant

Sworn to and Subscribed before Me
This *30th* Day of *April, 2006*

_____
Notary Public

KEEPING A MOTOR VEHICLE IN A STATE OF DISREPAIR—IPMC 302.8

STATE OF *ILLINOIS*
COUNTY OF DUPAGE
VILLAGE OF WOODRIDGE
v.
NAME: *ROBERT MEYER*
ADDRESS: *4617 Carousel St.*
CITY: *Woodridge, Illinois 60517*

The undersigned says that on or about *April 2, 2006*, at or about *3:00 p.m.* the Defendant did unlawfully commit the offense of **Keeping a Motor Vehicle In a State of Disrepair** in violation of *2006* **IPMC-302.8** as amended and adopted by reference in Section *8-1J-1(A)* of the *Village of Woodridge* Code; in that said Defendant, the owner of *4617 Carousel St., Woodridge, Illinois*, did maintain an inoperative motor vehicle on the premises in that *a red van in the driveway had no engine, and four flat tires.*

Joan Rogers
Complainant

Sworn to and Subscribed before Me
This *30th* Day of *April, 2006*

_____
Notary Public

DEFACEMENT OF PROPERTY—IPMC 302.9

STATE OF *ILLINOIS*
COUNTY OF DUPAGE
VILLAGE OF WOODRIDGE
v.
NAME: *TIM MCKENNA*
ADDRESS: *2206 E. Second St.*
CITY: *Woodridge, Illinois 60517*

The undersigned says that on or about *November 18, 2006*, at or about *11:00 p.m.* the Defendant did unlawfully commit the offense of **Defacement of Property** in violation of *2006 IPMC-302.9* as amended and adopted by reference in Section *8-1J-1(A)* of the *Village of Woodridge* Code; in that said Defendant willfully* damaged** the exterior surface of private*** property, being the *garage door of the residence* located at *4617 Carousel St., Woodridge, Illinois*, by *spraying**** graffiti on the door.*

Joan Rogers
Complainant

Sworn to and Subscribed before Me
This *19th* Day of *November, 2006*

_____
Notary Public

   *Or, wantonly.
  **Or, mutilated or defaced.
 ***Or, public.
****Or, by marking or carving.

SAMPLE COMPLAINT

FAILURE TO RESTORE DEFACED PROPERTY—IPMC 302.9

STATE OF *ILLINOIS*
COUNTY OF DUPAGE
VILLAGE OF WOODRIDGE
v.
NAME: *ROBERT MEYER*
ADDRESS: *4617 Carousel St.*
CITY: *Woodridge, Illinois 60517*

The undersigned says that on or about *December 15, 2006*, at or about *9:00 a.m.* the Defendant did unlawfully commit the offense of **Failure to Restore Defaced Property** in violation of ***2006 IPMC-302.9*** as amended and adopted by reference in Section *8-1J-1(A)* of the *Village of Woodridge* Code; in that said Defendant, the owner of *4617 Carousel St., Woodridge, Illinois*, failed to restore the exterior surface of a structure damaged* by graffiti** being *the garage door of the residence* located at *4617 Carousel St., Woodridge, Illinois,* to an approved state of maintenance and repair.

Joan Rogers
Complainant

Sworn to and Subscribed before Me
This *19th* Day of *December, 2006*

_____
Notary Public

 *Or, mutilated or defaced.
**Or, marking or carving.

IPMC 303   Swimming Pool, SPA, and Hot Tub Violations

SAMPLE COMPLAINT
- -

FAILURE TO MAINTAIN SWIMMING POOL—IPMC 303.1

STATE OF *ILLINOIS*
COUNTY OF DUPAGE
VILLAGE OF WOODRIDGE
v.
NAME: *ROBERT MEYER*
ADDRESS: *4617 Carousel St.*
CITY: *Woodridge, Illinois 60517*

The undersigned says that on or about *October 3, 2006,* at or about *3:00 p.m.* the Defendant did unlawfully commit the offense of **Failure to Maintain Swimming Pool** in violation of *2006* **IPMC-303.1** as amended and adopted by reference in Section *8-1J-1(A)* of the *Village of Woodridge* Code; in that said Defendant, the owner of *4617 Carousel St., Woodridge, Illinois,* failed to maintain the swimming pool upon such premises in a clean and sanitary condition and in good repair in that *the pool has a large crack in it and is filled with stagnant water with decaying leaves.*

Joan Rogers
Complainant

Sworn to and Subscribed before Me
This *28th* Day of *October, 2006*

_____
Notary Public

IMPROPER SWIMMING POOL ENCLOSURE—IPMC 303.2

STATE OF *ILLINOIS*
COUNTY OF DUPAGE
VILLAGE OF WOODRIDGE
v.
NAME: *ROBERT MEYER*
ADDRESS: *4617 Carousel St.*
CITY: *Woodridge, Illinois 60517*

The undersigned says that on or about *October 3, 2006*, at or about *3:00 p.m.* the Defendant did unlawfully commit the offense of **Improper Swimming Pool Enclosure** in violation of *2006* **IPMC-303.2** as amended and adopted by reference in Section *8-1J-1(A)* of the *Village of Woodridge* Code; in that said Defendant, the owner of *4617 Carousel St., Woodridge, Illinois,* failed to provide a fence or barrier of at least 48 inches in height above the finished ground level of the pool and which completely surrounds the swimming pool.*

Joan Rogers

Complainant

Sworn to and Subscribed before Me
This *28th* Day of *October, 2006*

_____

Notary Public

*Or, failed to provide a properly installed gate or door that is self-latching or did remove, replace, or change the pool enclosure in a manner that reduced the effectiveness of the safety barrier.

COMMENT: Spas or hot tubs with a safety cover that complies with ASTM F 1346-91 (2003) are exempt from the provisions of this section in the 2006 IPMC.

IPMC 304   Exterior Structure Violations

IPMC 304.18.1—Doors

IPMC 304.18.2—Windows

IPMC 304.18.3—Basement Hatchways

SAMPLE COMPLAINT

FAILURE TO MAINTAIN EXTERIOR STRUCTURE—IPMC 304.1

STATE OF *ILLINOIS*
COUNTY OF DUPAGE
VILLAGE OF WOODRIDGE
v.
NAME: *TREVOR BISHOP*
ADDRESS: *129 Meadow St.*
CITY: *Woodridge, Illinois 60517*

The undersigned says that on or about *May 24, 2006*, at or about *2:00 p.m.* the Defendant did unlawfully commit the offense of **Failure to Maintain Exterior Structure** in violation of *2006* **IPMC-304.1** as amended and adopted by reference in Section *8-1J-1(A)* of the *Village of Woodridge* Code; in that said Defendant, the owner of *129 Meadow St., Woodridge, Illinois,* failed to keep the exterior of the structure in good repair, structurally sound, and sanitary, so as not to pose a threat to the public health, safety, or welfare in that *the roof is collapsing, the wall on the east side has rotting clapboard siding, and there are broken windows on the first floor.*

Don Lay
Complainant

Sworn to and Subscribed before Me
This *17th* Day of *July, 2006*

_____
Notary Public

FAILURE TO MAINTAIN EXTERIOR SURFACES (WOOD SURFACE)—IPMC 304.2

STATE OF *ILLINOIS*
COUNTY OF DUPAGE
VILLAGE OF WOODRIDGE
v.
NAME: *TREVOR BISHOP*
ADDRESS: *129 Meadow St.*
CITY: *Woodridge, Illinois 60517*

The undersigned says that on or about *May 24, 2006*, at or about *2:00 p.m.* the Defendant did unlawfully commit the offense of **Failure to Maintain Exterior Surface** in violation of ***2006* IPMC-304.2** as amended and adopted by reference in Section *8-1J-1(A)* of the *Village of Woodridge* Code; in that said Defendant, the owner of *129 Meadow St., Woodridge, Illinois,* failed to maintain the exterior surface of the structure, being the porch,* in good condition in that *he failed to remove flaking, peeling, and chipped paint.*

Don Lay
Complainant

Sworn to and Subscribed before Me
This *17th* Day of *June, 2006*

_____
Notary Public

*Or, door, door and window frame, cornice, trim, balcony, deck, fence, or other exterior surface.

FAILURE TO MAINTAIN EXTERIOR SURFACES (SIDING AND MASONRY)—IPMC 304.2

STATE OF *ILLINOIS*
COUNTY OF DUPAGE
VILLAGE OF WOODRIDGE
v.
NAME: *TREVOR BISHOP*
ADDRESS: *129 Meadow St.*
CITY: *Woodridge, Illinois 60517*

The undersigned says that on or about *May 24, 2006*, at or about *2:00 p.m.* the Defendant did unlawfully commit the offense of **Failure to Maintain Exterior Surface** in violation of *2006* **IPMC-304.2** as amended and adopted by reference in Section *8-1J-1(A)* of the *Village of Woodridge* Code; in that said Defendant, the owner of *129 Meadow St., Woodridge, Illinois,* failed to maintain the exterior surface of the structure, being the siding* in a weather-resistant and watertight condition, in that *part of the siding is missing on the south side of the structure.*

Don Lay

Complainant

Sworn to and Subscribed before Me
This *17th* Day of *June, 2006*

_____

Notary Public

*Or, masonry joints.

FAILURE TO MAINTAIN EXTERIOR SURFACES (METAL)—IPMC 304.2

STATE OF *ILLINOIS*
COUNTY OF DUPAGE
VILLAGE OF WOODRIDGE
v.
NAME: *TREVOR BISHOP*
ADDRESS: *129 Meadow St.*
CITY: *Woodridge, Illinois 60517*

The undersigned says that on or about *May 24, 2006,* at or about *2:00 p.m.* the Defendant did unlawfully commit the offense of **Failure to Maintain Exterior Surface** in violation of *2006* **IPMC-304.2** as amended and adopted by reference in Section *8-1J-1(A)* of the *Village of Woodridge* Code; in that said Defendant, the owner of *129 Meadow St., Woodridge, Illinois,* failed to maintain the exterior surface of the structure, being metal surfaces, in that the defendant failed to coat the surface of the *metal porch railings* so as to inhibit rust and corrosion.*

Don Lay
Complainant

Sworn to and Subscribed before Me
This *17th* Day of *June, 2006*

_____
Notary Public

*Or, the defendant failed to remove oxidation stains.

COMMENT: This sample complaint is also suitable for violations of IPMC 304.9 and 304.11 for failing to maintain metal surfaces on overhang extensions or chimneys and towers.

SAMPLE COMPLAINT

- -

FAILURE TO POST PREMISES IDENTIFICATION—IPMC 304.3

STATE OF *ILLINOIS*
COUNTY OF DUPAGE
VILLAGE OF WOODRIDGE
v.
NAME: *TREVOR BISHOP*
ADDRESS: *129 Meadow St.*
CITY: *Woodridge, Illinois 60517*

The undersigned says that on or about *May 24, 2006,* at or about *2:00 p.m.* the Defendant did unlawfully commit the offense of **Failure to Post Premises Identification** in violation of **2006 IPMC-304.3** as amended and adopted by reference in Section *8-1J-1(A)* of the *Village of Woodridge* Code; in that said Defendant, the owner of *129 Meadow St., Woodridge, Illinois,* failed to have approved address numbers placed on the structure in a position to be plainly legible and visible from the street or road fronting the property.

Don Lay

Complainant

Sworn to and Subscribed before Me
This *17th* Day of *June, 2006*

_____

Notary Public

FAILURE TO MAINTAIN STRUCTURAL MEMBERS—IPMC 304.4

STATE OF *ILLINOIS*
COUNTY OF DUPAGE
VILLAGE OF WOODRIDGE
v.
NAME: *TREVOR BISHOP*
ADDRESS: *129 Meadow St.*
CITY: *Woodridge, Illinois 60517*

The undersigned says that on or about *May 24, 2006*, at or about *2:00 p.m.* the Defendant did unlawfully commit the offense of **Failure to Maintain Structural Members** in violation of *2006* **IPMC-304.4** as amended and adopted by reference in Section *8-1J-1(A)* of the *Village of Woodridge* Code; in that said Defendant, the owner of *129 Meadow St., Woodridge, Illinois,* failed to maintain a structural member on the premises in that *one of the support columns in the basement is rotten* and not capable of safely supporting the imposed dead and live loads of the structure.

Don Lay
Complainant

Sworn to and Subscribed before Me
This *17th Day* of *June, 2006*

_____
Notary Public

SAMPLE COMPLAINT

FAILURE TO MAINTAIN FOUNDATION—IPMC 304.5

STATE OF *ILLINOIS*
COUNTY OF DUPAGE
VILLAGE OF WOODRIDGE
v.
NAME: *TREVOR BISHOP*
ADDRESS: *129 Meadow St.*
CITY: *Woodridge, Illinois 60517*

The undersigned says that on or about *May 24, 2006*, at or about *2:00 p.m.* the Defendant did unlawfully commit the offense of **Failure to Maintain Foundation** in violation of *2006* **IPMC-304.5** as amended and adopted by reference in Section *8-1J-1(A)* of the *Village of Woodridge* Code; in that said Defendant, the owner of *129 Meadow St., Woodridge, Illinois,* failed to maintain the foundation of the structure in that there are cracks and breaks *in the north foundation wall* so as to allow the entry of rodents and other pests.

Don Lay
Complainant

Sworn to and Subscribed before Me
This *17th* Day of *June, 2006*

_____

Notary Public

**FAILURE TO MAINTAIN EXTERIOR WALLS (HOLES, BREAKS, AND LOOSE OR ROTTING MATERIAL)—
IPMC 304.6**

STATE OF *ILLINOIS*
COUNTY OF DUPAGE
VILLAGE OF WOODRIDGE
v.
NAME: *TREVOR BISHOP*
ADDRESS: *129 Meadow St.*
CITY: *Woodridge, Illinois 60517*

The undersigned says that on or about *May 24, 2006,* at or about *2:00 p.m.* the Defendant did unlawfully commit the offense of **Failure to Maintain Exterior Walls** in violation of **2006 IPMC-304.6** as amended and adopted by reference in Section *8-1J-1(A)* of the *Village of Woodridge* Code; in that said Defendant, the owner of *129 Meadow St., Woodridge, Illinois,* failed to maintain the exterior walls of the structure in that the exterior walls are not free from holes, breaks, and loose or rotting materials, *being missing siding on the south wall of the structure.*

Don Lay

Complainant

Sworn to and Subscribed before Me
This *17th* Day of *June, 2006*

_____
Notary Public

FAILURE TO MAINTAIN EXTERIOR WALLS (SURFACE COATING)—IPMC 304.6

STATE OF *ILLINOIS*
COUNTY OF DUPAGE
VILLAGE OF WOODRIDGE
v.
NAME: *TREVOR BISHOP*
ADDRESS: *129 Meadow St.*
CITY: *Woodridge, Illinois 60517*

The undersigned says that on or about *May 24, 2006*, at or about *2:00 p.m.* the Defendant did unlawfully commit the offense of **Failure to Maintain Exterior Walls** in violation of **2006 IPMC-304.6** as amended and adopted by reference in Section *8-1J-1(A)* of the *Village of Woodridge* Code; in that said Defendant, the owner of *129 Meadow St., Woodridge, Illinois,* failed to maintain the exterior walls of the structure in that *the exterior soffit and fascia boards are not coated with a non-permeable substance on the exterior of the house** so as to prevent deterioration.

Don Lay

Complainant

Sworn to and Subscribed before Me
This *17th* Day of *June, 2006*

_____

Notary Public

SAMPLE COMPLAINT

- -

FAILURE TO MAINTAIN ROOF AND FLASHING—IPMC 304.7

STATE OF *ILLINOIS*
COUNTY OF DUPAGE
VILLAGE OF WOODRIDGE
v.
NAME: *TREVOR BISHOP*
ADDRESS: *129 Meadow St.*
CITY: *Woodridge, Illinois 60517*

The undersigned says that on or about *May 24, 2006,* at or about *2:00 p.m.* the Defendant did unlawfully commit the offense of **Failure to Maintain Roof and Flashing** in violation of **2006 IPMC-304.7** as amended and adopted by reference in Section *8-1J-1(A)* of the *Village of Woodridge* Code; in that said Defendant, the owner of *129 Meadow St., Woodridge, Illinois,* failed to maintain the roof and flashing of the structure in a condition so as to be sound, tight and not have defects that admit rain in that *there are numerous shingles missing on the roof and the flashing is not tight where it meets the roof.*

Don Lay
Complainant

Sworn to and Subscribed before Me
This *17th* Day of *March, 2006*

_____
Notary Public

FAILURE TO MAINTAIN ROOF DRAINAGE—IPMC 304.7

STATE OF *ILLINOIS*
COUNTY OF DUPAGE
VILLAGE OF WOODRIDGE
v.
NAME: *TREVOR BISHOP*
ADDRESS: *129 Meadow St.*
CITY: *Woodridge, Illinois 60517*

The undersigned says that on or about *May 24, 2006*, at or about *2:00 p.m.* the Defendant did unlawfully commit the offense of **Failure to Maintain Roof Drainage** in violation of *2006* **IPMC-304.7** as amended and adopted by reference in Section *8-1J-1(A)* of the *Village of Woodridge* Code; in that said Defendant, the owner of *129 Meadow St., Woodridge, Illinois,* failed to maintain the roof drainage of the structure in a condition so as to prevent dampness or deterioration in the walls or interior portion of the structure in that *there are no gutters on the structure.*

Don Lay
Complainant

Sworn to and Subscribed before Me
This *17th* Day of *June, 2006*

_____

Notary Public

FAILURE TO MAINTAIN ROOF DRAINS, GUTTERS, AND DOWNSPOUTS—IPMC 304.7

STATE OF *ILLINOIS*
COUNTY OF DUPAGE
VILLAGE OF WOODRIDGE
v.
NAME: *TREVOR BISHOP*
ADDRESS: *129 Meadow St.*
CITY: *Woodridge, Illinois 60517*

The undersigned says that on or about *May 24, 2006*, at or about *2:00 p.m.* the Defendant did unlawfully commit the offense of **Failure to Maintain Roof Drains, Gutters, and Downspouts** in violation of *2006* **IPMC-304.7** as amended and adopted by reference in Section *8-1J-1(A)* of the *Village of Woodridge* Code; in that said Defendant, the owner of *129 Meadow St., Woodridge, Illinois,* failed to maintain the roof drains, gutters, and downspouts of the structure in good repair and free from obstructions in that *the drains, gutters, and downspouts are clogged with debris.*

Don Lay

Complainant

Sworn to and Subscribed before Me
This *17th* Day of *June, 2006*

_____
Notary Public

SAMPLE COMPLAINT
- -

CREATING A PUBLIC NUISANCE (ROOF WATER)—IPMC 304.7

STATE OF *ILLINOIS*
COUNTY OF DUPAGE
VILLAGE OF WOODRIDGE
v.
NAME: *TREVOR BISHOP*
ADDRESS: *129 Meadow St.*
CITY: *Woodridge, Illinois 60517*

The undersigned says that on or about *February 24, 2006,* at or about *2:00 p.m.* the Defendant did unlawfully commit the offense of **Creating a Public Nuisance** in violation of *2006* **IPMC-304.7** as amended and adopted by reference in Section *8-1J-1(A)* of the *Village of Woodridge* Code; in that said Defendant, the owner of *129 Meadow St., Woodridge, Illinois,* created a public nuisance in that the roof water from the structure *is being discharged toward the public sidewalk and street where it is freezing* and creating a hazard to the public.

Don Lay
Complainant

Sworn to and Subscribed before Me
This *17th* Day of *March, 2006*

_____
Notary Public

FAILURE TO MAINTAIN DECORATIVE FEATURES—IPMC 304.8

STATE OF *ILLINOIS*
COUNTY OF DUPAGE
VILLAGE OF WOODRIDGE
v.
NAME: *LARRY SALG*
ADDRESS: *3859 Schiller Dr.*
CITY: *Woodridge, Illinois 60517*

The undersigned says that on or about *May 24, 2006*, at or about *2:00 p.m.* the Defendant did unlawfully commit the offense of **Failure to Maintain Decorative Features** in violation of ***2006* IPMC-304.8** as amended and adopted by reference in Section *8-1J-1(A)* of the *Village of Woodridge* Code; in that said Defendant, the owner of *34 West Reber St., Woodridge, Illinois,* failed to maintain the terra cotta trim* of the structure in good repair with proper anchorage and in a safe condition *in that the terra cotta trim is falling off the building, presenting a hazard to pedestrians.*

Don Lay

Complainant

Sworn to and Subscribed before Me
This *17th* Day of *June, 2006*

_____

Notary Public

*Or, cornices, belt courses, corbels, wall facings, or similar decorative features.

- -

FAILURE TO MAINTAIN OVERHANG EXTENSIONS—IPMC 304.9

STATE OF *ILLINOIS*
COUNTY OF DUPAGE
VILLAGE OF WOODRIDGE
v.
NAME: *LARRY SALG*
ADDRESS: *3859 Schiller Dr.*
CITY: *Woodridge, Illinois 60517*

The undersigned says that on or about *May 24, 2006*, at or about *2:00 p.m.* the Defendant did unlawfully commit the offense of **Failure to Maintain Overhang Extensions** in violation of **2006 IPMC-304.9** as amended and adopted by reference in Section *8-1J-1(A)* of the *Village of Woodridge* Code; in that said Defendant, the owner of *34 West Reber St., Woodridge, Illinois,* failed to maintain the canopy* on the front of the structure in good repair with proper anchorage and in a safe condition *in that part of the canopy has pulled away from the exterior wall and is hanging loose, and there are holes in the canopy material.*

Don Lay
Complainant

Sworn to and Subscribed before Me
This *17th* Day of *June, 2006*

_____
Notary Public

*Or, marquee, sign, metal awning, fire escape, standpipe, or exhaust duct.

- -

FAILURE TO MAINTAIN EXTERIOR STAIRWAY*—IPMC 304.10

STATE OF *ILLINOIS*
COUNTY OF DUPAGE
VILLAGE OF WOODRIDGE
v.
NAME: *TREVOR BISHOP*
ADDRESS: *129 Meadow St.*
CITY: *Woodridge, Illinois 60517*

The undersigned says that on or about *May 24, 2006*, at or about *2:00 p.m.* the Defendant did unlawfully commit the offense of **Failure to Maintain Exterior Stairway*** in violation of ***2006* IPMC-304.10** as amended and adopted by reference in Section *8-1J-1(A)* of the Village of Woodridge Code; in that said Defendant, the owner of *129 Meadow St., Woodridge, Illinois,* failed to maintain the stairway* of the structure in a structurally sound manner, in good repair with proper anchorage and capable of supporting the imposed loads in that *the treads and risers on the staircase on the east side of the structure are rotten.*

Don Lay

Complainant

Sworn to and Subscribed before Me
This *17th Day* of *June, 2006*

_____

Notary Public

*Or, deck, porch, balcony, or all appurtenances attached thereto.

SAMPLE COMPLAINT

FAILURE TO MAINTAIN CHIMNEYS AND TOWERS—IPMC 304.11

STATE OF *ILLINOIS*
COUNTY OF DUPAGE
VILLAGE OF WOODRIDGE
v.
NAME: *TREVOR BISHOP*
ADDRESS: *129 Meadow St.*
CITY: *Woodridge, Illinois 60517*

The undersigned says that on or about *May 24, 2006*, at or about *2:00 p.m.* the Defendant did unlawfully commit the offense of **Failure to Maintain Chimney*** in violation of *2006* **IPMC-304.11** as amended and adopted by reference in Section *8-1J-1(A)* of the Village of Woodridge Code; in that said Defendant, the owner of *129 Meadow St., Woodridge, Illinois,* failed to maintain the chimney* on the structure in a structurally safe and sound manner, and in good repair in that *bricks are missing and loose, and the chimney needs tuck pointing.*

Don Lay

Complainant

Sworn to and Subscribed before Me
This *17th* Day of *June, 2006*

_____

Notary Public

*Or, cooling towers, smokestacks, or similar appurtenances.

FAILURE TO MAINTAIN HANDRAILS AND GUARDS—IPMC 304.12

STATE OF *ILLINOIS*
COUNTY OF DUPAGE
VILLAGE OF WOODRIDGE
v.
NAME: *TREVOR BISHOP*
ADDRESS: *129 Meadow St.*
CITY: *Woodridge, Illinois 60517*

The undersigned says that on or about *May 24, 2006*, at or about *2:00 p.m.* the Defendant did unlawfully commit the offense of **Failure to Maintain Handrail*** in violation of *2006* **IPMC-304.12** as amended and adopted by reference in Section *8-1J-1(A)* of the Village of Woodridge Code; in that said Defendant, the owner of *129 Meadow St., Woodridge, Illinois,* failed to maintain the handrail in good condition, firmly fastened, and capable of supporting normally imposed loads in that *the handrail is not attached to the newel post.*

Don Lay
Complainant

Sworn to and Subscribed before Me
This *17th* Day of *June, 2006*

_____
Notary Public

*Or, guard.

FAILURE TO MAINTAIN WINDOW, SKYLIGHT AND DOOR FRAMES—IPMC 304.13

STATE OF *ILLINOIS*
COUNTY OF DUPAGE
VILLAGE OF WOODRIDGE
v.
NAME: *TREVOR BISHOP*
ADDRESS: *129 Meadow St.*
CITY: *Woodridge, Illinois 60517*

The undersigned says that on or about *May 24, 2006*, at or about *2:00 p.m.* the Defendant did unlawfully commit the offense of **Failure to Maintain Window*** in violation of *2006 IPMC-***304.13** as amended and adopted by reference in Section *8-1J-1(A)* of the Village of Woodridge Code; in that said Defendant, the owner of *129 Meadow St., Woodridge, Illinois,* failed to maintain a window in sound condition, good repair, and weather tight in that *a window on the west side of the structure on the first floor is cracked and allows water to penetrate into the interior.*

Don Lay
Complainant

Sworn to and Subscribed before Me
This *17th* Day of *June, 2006*

_____
Notary Public

*Or, skylight or door frame.

FAILURE TO MAINTAIN GLAZING MATERIAL—IPMC 304.13.1

STATE OF *ILLINOIS*
COUNTY OF DUPAGE
VILLAGE OF WOODRIDGE
v.
NAME: *TREVOR BISHOP*
ADDRESS: *129 Meadow St.*
CITY: *Woodridge, Illinois 60517*

The undersigned says that on or about *May 24, 2006,* at or about *2:00 p.m.* the Defendant did unlawfully commit the offense of **Failure to Maintain Glazing Material** in violation of ***2006 IPMC-304.13.1*** as amended and adopted by reference in Section *8-1J-1(A)* of the Village of Woodridge Code; in that said Defendant, the owner of *129 Meadow St., Woodridge, Illinois,* failed to maintain the glazing material *in the windows* free from cracks and holes in that *the windows on the west side of the structure on all floors have cracked and missing glazing.*

Don Lay
Complainant

Sworn to and Subscribed before Me
This *17th* Day of *June, 2006*

_____
Notary Public

SAMPLE COMPLAINT

FAILURE TO PROVIDE OPENABLE WINDOWS—IPMC 304.13.2

STATE OF *ILLINOIS*
COUNTY OF DUPAGE
VILLAGE OF WOODRIDGE

v.

NAME: *TREVOR BISHOP*
ADDRESS: *129 Meadow St.*
CITY: *Woodridge, Illinois 60517*

The undersigned says that on or about *May 24, 2006*, at or about *2:00 p.m.* the Defendant did unlawfully commit the offense of **Failure to Provide Openable Windows** in violation of **2006 IPMC-304.13.2** as amended and adopted by reference in Section *8-1J-1(A)* of the Village of Woodridge Code; in that said Defendant, the owner of *129 Meadow St., Woodridge, Illinois,* failed to provide windows capable of being easily openable and capable of being held in position by window hardware in that *the windows on the south side of the structure cannot open because they have been painted shut.*

Don Lay
Complainant

Sworn to and Subscribed before Me
This *17th* Day of *June, 2006*

_____
Notary Public

- -

FAILURE TO SUPPLY INSECT SCREENS—IPMC 304.14

STATE OF *ILLINOIS*

COUNTY OF DUPAGE

VILLAGE OF WOODRIDGE

v.

NAME: *TREVOR BISHOP*

ADDRESS: *129 Meadow St.*

CITY: *Woodridge, Illinois 60517*

The undersigned says that on or about *May 24, 2006*, at or about *2:00 p.m.* the Defendant did unlawfully commit the offense of **Failure to Provide Insect Screens** in violation of *2006 IPMC-304.14* as amended and adopted by reference in Section *8-1J-1(A)* of the Village of Woodridge Code; in that said Defendant, the owner of *129 Meadow St., Woodridge, Illinois,* failed to provide approved tightly fitting screens for the windows* in the bedrooms and kitchen** of the premises.

Don Lay

Complainant

Sworn to and Subscribed before Me

This *17th* Day of *June, 2006*

_____

Notary Public

 *Or, door or other outside opening.

**Or, other habitable room, food preparation area, food service area, or any area where products to be included or utilized in food for human consumption are processed, manufactured, packaged, or stored.

COMMENT: Screens must be not less than 16 mesh per inch (16 mesh per 25 mm). The date of the offense must be during the time set by the ordinance when screens must be provided, e.g., April 1 to December 1. Screens are not required where other approved means are used, e.g., air curtains or insect repellent fans.

SAMPLE COMPLAINT

- -

FAILURE TO MAINTAIN SWINGING DOOR—2003 IPMC 304.14

STATE OF *ILLINOIS*
COUNTY OF DUPAGE
VILLAGE OF WOODRIDGE
v.
NAME: *TREVOR BISHOP*
ADDRESS: *129 Meadow St.*
CITY: *Woodridge, Illinois 60517*

The undersigned says that on or about *May 24, 2006,* at or about *2:00 p.m.* the Defendant did unlawfully commit the offense of **Failure to Maintain Swinging Door** in violation of *2003* **IPMC-304.14** as amended and adopted by reference in Section *8-1J-1(A)* of the Village of Woodridge Code; in that said Defendant, the owner of *129 Meadow St., Woodridge, Illinois,* failed to maintain the swinging door used for *on the back porch* in good working condition *in that the self-closing device, the pneumatic door closer, is broken and inoperable.*

Don Lay
Complainant

Sworn to and Subscribed before Me
This *17th* Day of *June, 2006*

_____
Notary Public

FAILURE TO MAINTAIN SWINGING DOOR—2006 IPMC 304.14

STATE OF *ILLINOIS*
COUNTY OF DUPAGE
VILLAGE OF WOODRIDGE
v.
NAME: *TREVOR BISHOP*
ADDRESS: *129 Meadow St.*
CITY: *Woodridge, Illinois 60517*

The undersigned says that on or about *May 24, 2006*, at or about *2:00 p.m.* the Defendant did unlawfully commit the offense of **Failure to Maintain Swinging Door** in violation of ***2006 IPMC-304.14*** as amended and adopted by reference in Section *8-1J-1(A)* of the Village of Woodridge Code; in that said Defendant, the owner of *129 Meadow St., Woodridge, Illinois,* failed to maintain the swinging screen door used for insect control* *on the back porch* in good working condition *in that the self-closing device, the pneumatic door closer, is broken and inoperable.*

Don Lay
Complainant

Sworn to and Subscribed before Me
This *17th* Day of *June, 2006*

_____
Notary Public

*In the 2006 IPMC, the self-closing requirement is only applicable to screen doors used for insect control.

SAMPLE COMPLAINT
- -

FAILURE TO MAINTAIN EXTERIOR DOOR—IPMC 304.15

STATE OF *ILLINOIS*
COUNTY OF DUPAGE
VILLAGE OF WOODRIDGE
v.
NAME: *TREVOR BISHOP*
ADDRESS: *129 Meadow St.*
CITY: *Woodridge, Illinois 60517*

The undersigned says that on or about *May 24, 2006*, at or about *2:00 p.m.* the Defendant did unlawfully commit the offense of **Failure to Maintain Exterior Door*** in violation of *2006* **IPMC-304.15** as amended and adopted by reference in Section *8-1J-1(A)* of the *Village of Woodridge* Code; in that said Defendant, the owner of *129 Meadow St., Woodridge, Illinois,* failed to maintain the exterior door* *on the back porch* in good condition *in that the door is warped and does not close completely.*

Don Lay

Complainant

Sworn to and Subscribed before Me
This *17th* Day of *June, 2006*

_____
Notary Public

*Or, door assembly or hardware.

IMPROPER LOCK—IPMC 304.15

STATE OF *ILLINOIS*
COUNTY OF DUPAGE
VILLAGE OF WOODRIDGE
v.
NAME: *TREVOR BISHOP*
ADDRESS: *129 Meadow St.*
CITY: *Woodridge, Illinois 60517*

The undersigned says that on or about *May 24, 2006,* at or about *2:00 p.m.* the Defendant did unlawfully commit the offense of **Improper Lock** in violation of *2006* **IPMC-304.15** as amended and adopted by reference in Section *8-1J-1(A)* of the *Village of Woodridge* Code; in that said Defendant, the owner of *129 Meadow St., Woodridge, Illinois,* did allow an improper lock on the door *on the back porch* in that *a key is needed to unlock the door to gain access to the exterior* which is forbidden by Section 702.3 of the *2006* IPMC as amended and adopted by reference in Section *8-1J-1(A)* of the *Village of Woodridge* Code.

Don Lay
Complainant

Sworn to and Subscribed before Me
This *17th* Day of *June, 2006*

_____
Notary Public

- -

FAILURE TO MAINTAIN BASEMENT HATCHWAY—IPMC 304.16

STATE OF *ILLINOIS*
COUNTY OF DUPAGE
VILLAGE OF WOODRIDGE
v.
NAME: *TREVOR BISHOP*
ADDRESS: *129 Meadow St.*
CITY: *Woodridge, Illinois 60517*

The undersigned says that on or about *May 24, 2006,* at or about *2:00 p.m.* the Defendant did unlawfully commit the offense of **Failure to Maintain Basement Hatchway** in violation of **2006 IPMC-304.16** as amended and adopted by reference in Section *8-1J-1(A)* of the *Village of Woodridge* Code; in that said Defendant, the owner of *129 Meadow St., Woodridge, Illinois,* failed to maintain the basement hatchway so as to prevent the entrance of rodents, rain, and surface drainage water in that *the hatchway door has cracks and holes in it and does not close completely* .

Don Lay
Complainant

Sworn to and Subscribed before Me
This *17th* Day of *June, 2006*

_____
Notary Public

FAILURE TO SUPPLY BASEMENT WINDOW GUARD—IPMC 304.17

STATE OF *ILLINOIS*
COUNTY OF DUPAGE
VILLAGE OF WOODRIDGE
v.
NAME: *TREVOR BISHOP*
ADDRESS: *129 Meadow St.*
CITY: *Woodridge, Illinois 60517*

The undersigned says that on or about *May 24, 2006*, at or about *2:00 p.m.* the Defendant did unlawfully commit the offense of **Failure to Supply Basement Window Guard** in violation of *2006* **IPMC-304.17** as amended and adopted by reference in Section *8-1J-1(A)* of the *Village of Woodridge* Code; in that said Defendant, the owner of *129 Meadow St., Woodridge, Illinois,* failed to supply a storm window* for the openable basement window *on the west side of the structure* so as to prevent the entrance of rodents.

Don Lay
_____
Complainant

Sworn to and Subscribed before Me
This *17th* Day of *June, 2006*

_____
Notary Public

*Or, rodent shield or other approved protection.

SAMPLE COMPLAINT

FAILURE TO PROVIDE BUILDING SECURITY—IPMC 304.18

STATE OF *ILLINOIS*
COUNTY OF DUPAGE
VILLAGE OF WOODRIDGE
v.
NAME: *LARRY SALG*
ADDRESS: *3891 Schiller Dr.*
CITY: *Woodridge, Illinois 60517*

The undersigned says that on or about *June 17, 2006,* at or about *2:00 p.m.* the Defendant did unlawfully commit the offense of **Failure to Provide Building Security** in violation of *2006 IPMC-304.18 as amended and adopted by reference in Section 8-1J-1(A) of the Village* of Woodridge Code; in that said Defendant, the owner of *34 West Reber St., Woodridge, Illinois,* failed to provide security for the dwelling units* in the building in that *there are no locks for any of the doors** so as to provide security for the occupants and property within.

Don Lay

Complainant

Sworn to and Subscribed before Me
This *8th* Day of *July, 2006*

_____

Notary Public

  *Or, room units or housekeeping units.
**Or, windows or hatchways.

FAILURE TO PROVIDE DEADBOLT LOCK—IPMC 304.18.1

STATE OF *ILLINOIS*
COUNTY OF DUPAGE
VILLAGE OF WOODRIDGE
v.
NAME: *LARRY SALG*
ADDRESS: *3891 Schiller Dr.*
CITY: *Woodridge, Illinois 60517*

The undersigned says that on or about *June 17, 2006*, at or about *2:00 p.m.* the Defendant did unlawfully commit the offense of **Failure to Provide Deadbolt Lock** in violation of ***2006 IPMC-304.18.1*** as amended and adopted by reference in Section *8-1J-1(A)* of the *Village of Woodridge* Code; in that said Defendant, the owner of *34 West Reber St., Woodridge, Illinois,* failed to provide deadbolt locks on the doors of the dwelling units* rented** in the building in that *there are no deadbolt locks for any of the doors.*

Don Lay
Complainant

Sworn to and Subscribed before Me
This *8th* Day of *July, 2006*

_____

Notary Public

  *Or, rooming units or housekeeping units.
**Or, leased or let.

SAMPLE COMPLAINT

FAILURE TO PROVIDE PROPER DEADBOLT LOCK—IPMC 304.18.1

STATE OF *ILLINOIS*
COUNTY OF DUPAGE
VILLAGE OF WOODRIDGE
v.
NAME: *LARRY SALG*
ADDRESS: *3891 Schiller Dr.*
CITY: *Woodridge, Illinois 60517*

The undersigned says that on or about *June 19, 2006,* at or about *2:00 p.m.* the Defendant did unlawfully commit the offense of **Failure to Provide Proper Deadbolt Lock** in violation of ***2006* IPMC-304.18.1** as amended and adopted by reference in Section *8-1J-1(A)* of the *Village of Woodridge* Code; in that said Defendant, the owner of *34 West Reber St., Woodridge, Illinois,* failed to provide a proper deadbolt lock on the door of a dwelling unit* rented** in the building in that said deadbolt is not readily openable from the side from which egress is to be made without the need for a tool, combination of a key and tool, or special knowledge or effort to be opened.***

Complainant

Sworn to and Subscribed before Me
This *8th* Day of *July, 2006*

_____

Notary Public

   *Or, rooming units or housekeeping units.

  **Or, leased or let.

***Or, said deadbolt has a lock throw of less than 1 inch (25 mm) or the deadbolt was not installed according to the manufacturer's specification, or the deadbolt was not maintained in good working order, or the lock is a sliding bolt, not a deadbolt.

FAILURE TO PROVIDE WINDOW SASH LOCKING DEVICES—IPMC 304.18.2

STATE OF *ILLINOIS*
COUNTY OF DUPAGE
VILLAGE OF WOODRIDGE
v.
NAME: *LARRY SALG*
ADDRESS: *3891 Schiller Dr.*
CITY: *Woodridge, Illinois 60517*

The undersigned says that on or about *June 17, 2006,* at or about *2:00 p.m.* the Defendant did unlawfully commit the offense of **Failure to Provide Window Sash Locking Devices** in violation of *2006* **IPMC-304.18.2** as amended and adopted by reference in Section *8-1J-1(A)* of the *Village of Woodridge* Code; in that said Defendant, the owner of *34 West Reber St., Woodridge, Illinois,* failed to provide window sash locking devices on the operable windows located within 6 feet above the ground level* of a dwelling unit,** *Unit 1D,* rented*** in the building.

Don Lay

Complainant

Sworn to and Subscribed before Me
This *8th* Day of *July, 2006*

_____

Notary Public

   *Or, a walking surface below that provides access to a dwelling unit.
  **Or, rooming units or housekeeping units.
***Or, leased or let.

- -

FAILURE TO PROVIDE BASEMENT HATCHWAY SECURITY DEVICE—IPMC 304.18.3

STATE OF *ILLINOIS*
COUNTY OF DUPAGE
VILLAGE OF WOODRIDGE
v.
NAME: *LARRY SALG*
ADDRESS: *3891 Schiller Dr.*
CITY: *Woodridge, Illinois 60517*

The undersigned says that on or about *June 17, 2006,* at or about *2:00 p.m.* the Defendant did unlawfully commit the offense of **Failure to Provide Basement Hatchway Security Device** in violation of ***2006* IPMC-304.18.3** as amended and adopted by reference in Section *8-1J-1(A)* of the *Village of Woodridge* Code; in that said Defendant, the owner of *34 West Reber St., Woodridge, Illinois,* failed to provide a device on the basement hatchway that provides access to the dwelling units* rented** in the building so as to prevent unauthorized entry therein.

Don Lay

Complainant

Sworn to and Subscribed before Me
This *8th* Day of *July, 2006*

_____
Notary Public

  *Or, rooming units or housekeeping units.
**Or, leased or let.

IPMC 305   Interior Structure Violations

SAMPLE COMPLAINT
- -

FAILURE TO MAINTAIN INTERIOR STRUCTURE—IPMC 305.1

STATE OF *ILLINOIS*
COUNTY OF DUPAGE
VILLAGE OF WOODRIDGE
v.
NAME: *TIM HALIK*
ADDRESS: *6560 Hollywood Blvd.*
CITY: *Woodridge, Illinois 60517*

The undersigned says that on or about *July 18, 2006*, at or about *2:00 p.m.* the Defendant did unlawfully commit the offense of **Failure to Maintain Interior Structure** in violation of **2006 IPMC-305.1** as amended and adopted by reference in Section *8-1J-1(A)* of the *Village of Woodridge* Code; in that said Defendant, the owner* of *6560 Hollywood Blvd., Woodridge, Illinois,* failed to keep the interior of the structure in good repair, structurally sound, and in a sanitary condition in that *the rafters in the attic are water damaged and rotten, there are holes in the living room walls, and the kitchen is filled with moldy dishes and food.*

Chuck Schmidt
Complainant

Sworn to and Subscribed before Me
This *28th* Day of *August, 2006*

_____
Notary Public

*Or, occupant.

COMMENT: An occupant is responsible for areas he or she occupies or controls. An owner is responsible for the shared or public areas of a structure containing a rooming house, housekeeping units, a hotel, a dormitory, two or more dwelling units, or two or more nonresidential occupancies and the exterior property.

FAILURE TO MAINTAIN INTERIOR STRUCTURAL MEMBERS—IPMC 305.2

STATE OF *ILLINOIS*
COUNTY OF DUPAGE
VILLAGE OF WOODRIDGE
v.
NAME: *TIM HALIK*
ADDRESS: *6560 Hollywood Blvd.*
CITY: *Woodridge, Illinois 60517*

The undersigned says that on or about *July 18, 2006*, at or about *2:00 p.m.* the Defendant did unlawfully commit the offense of **Failure to Maintain Interior Structural Members** in violation of *2006* **IPMC-305.2** as amended and adopted by reference in Section *8-1J-1(A)* of the *Village of Woodridge* Code; in that said Defendant, the owner of *6560 Hollywood Blvd., Woodridge, Illinois,* failed to keep the interior structural members structurally sound, capable of supporting the imposed loads, in that *the rafters in the attic are water damaged and rotten.*

Chuck Schmidt
Complainant

Sworn to and Subscribed before Me
This *28th* Day of *August, 2006*

_____
Notary Public

SAMPLE COMPLAINT

FAILURE TO MAINTAIN INTERIOR SURFACES—IPMC 305.3

STATE OF *ILLINOIS*
COUNTY OF DUPAGE
VILLAGE OF WOODRIDGE
v.
NAME: *TIM HALIK*
ADDRESS: *6560 Hollywood Blvd.*
CITY: *Woodridge, Illinois 60517*

The undersigned says that on or about *July 18, 2006*, at or about *2:00 p.m.* the Defendant did unlawfully commit the offense of **Failure to Maintain Interior Surfaces** in violation of **2006** **IPMC-305.3** as amended and adopted by reference in Section *8-1J-1(A)* of the *Village of Woodridge* Code; in that said Defendant, the owner of *6560 Hollywood Blvd., Woodridge, Illinois,* failed to keep the interior surfaces in a good, clean, and sanitary condition in that paint is peeling* *in the bedrooms and the bathroom ceiling is mildewed.*

Chuck Schmidt
Complainant

Sworn to and Subscribed before Me
This *28th* Day of *August, 2006*

_____
Notary Public

*Or, chipping, flaking, or abraded. Or, plaster is cracked or loose, or there is decayed wood or other defective surface conditions.

FAILURE TO MAINTAIN STAIRS AND WALKING SURFACES—IPMC 305.4

STATE OF *ILLINOIS*
COUNTY OF DUPAGE
VILLAGE OF WOODRIDGE
v.
NAME: *TIM HALIK*
ADDRESS: *6560 Hollywood Blvd.*
CITY: *Woodridge, Illinois 60517*

The undersigned says that on or about *July 18, 2006,* at or about *2:00 p.m.* the Defendant did unlawfully commit the offense of **Failure to Maintain Stairs and Walking Surfaces** in violation of *2006* **IPMC-305.4** as amended and adopted by reference in Section *8-1J-1(A)* of the *Village of Woodridge* Code; in that said Defendant, the owner of *6560 Hollywood Blvd., Woodridge, Illinois,* failed to keep the stairs* in sound condition and good repair in that *the staircase leading to the second floor has stairs that are warped and loose.*

Chuck Schmidt
Complainant

Sworn to and Subscribed before Me
This *28th* Day of *August, 2006*

_____

Notary Public

*Or, ramp, landing, balcony, porch, deck, or other walking surface.

SAMPLE COMPLAINT

FAILURE TO MAINTAIN HANDRAILS AND GUARDS—IPMC 305.5

STATE OF *ILLINOIS*
COUNTY OF DUPAGE
VILLAGE OF WOODRIDGE
v.
NAME: *TIM HALIK*
ADDRESS: *6560 Hollywood Blvd.*
CITY: *Woodridge, Illinois 60517*

The undersigned says that on or about *July 18, 2006,* at or about *2:00 p.m.* the Defendant did unlawfully commit the offense of **Failure to Maintain Handrails and Guards** in violation of *2006* **IPMC-305.5** as amended and adopted by reference in Section *8-1J-1(A)* of the *Village of Woodridge* Code; in that said Defendant, the owner of *6560 Hollywood Blvd., Woodridge, Illinois,* failed to maintain a handrail* in good condition, firmly fastened and capable of supporting normally imposed loads in that *the handrail leading to the basement is not fastened to the wall at the top of the stairs.*

Chuck Schmidt
Complainant

Sworn to and Subscribed before Me
This *28th* Day of *August, 2006*

_____
Notary Public

*Or, guard.

FAILURE TO MAINTAIN INTERIOR DOORS—IPMC 305.6

STATE OF *ILLINOIS*
COUNTY OF DUPAGE
VILLAGE OF WOODRIDGE
v.
NAME: *TIM HALIK*
ADDRESS: *6560 Hollywood Blvd.*
CITY: *Woodridge, Illinois 60517*

The undersigned says that on or about *July 18, 2006*, at or about *2:00 p.m.* the Defendant did unlawfully commit the offense of **Failure to Maintain Interior Doors** in violation of *2006* **IPMC-305.6** as amended and adopted by reference in Section *8-1J-1(A)* of the *Village of Woodridge* Code; in that said Defendant, the owner of *6560 Hollywood Blvd., Woodridge, Illinois,* failed to maintain an interior door because it is not properly and securely attached to the jamb* in that *the door to the bedroom cannot be closed because the hinge on the bottom is loose and improperly mounted.*

Chuck Schmidt

Complainant

Sworn to and Subscribed before Me
This *28th* Day of *August, 2006*

_____
Notary Public

*Or, header or track as intended by the manufacturer of the attachment hardware.

IPMC 306   Handrail and Guardrail Violations

IPMC 306.1—General

FAILURE TO MAINTAIN PROVIDE HANDRAILS—IPMC 306.1

STATE OF *ILLINOIS*
COUNTY OF DUPAGE
VILLAGE OF WOODRIDGE
v.
NAME: *TIM HALIK*
ADDRESS: *6560 Hollywood Blvd.*
CITY: *Woodridge, Illinois 60517*

The undersigned says that on or about *July 18, 2006*, at or about *2:00 p.m.* the Defendant did unlawfully commit the offense of **Failure to Provide Handrails** in violation of ***2006 IPMC-306.1*** as amended and adopted by reference in Section *8-1J-1(A)* of the *Village of Woodridge* Code; in that said Defendant, the owner of *6560 Hollywood Blvd., Woodridge, Illinois,* failed to provide handrails on an exterior* flight of stairs having more than four risers in that *the back flight of stairs does not have a handrail.*

Chuck Schmidt
Complainant

Sworn to and Subscribed before Me
This *28th* Day of *August, 2006*

_____
Notary Public

*Or, interior.

SAMPLE COMPLAINT

FAILURE TO MAINTAIN PROVIDE GUARDS—IPMC 306.1

STATE OF *ILLINOIS*
COUNTY OF DUPAGE
VILLAGE OF WOODRIDGE
v.
NAME: *TIM HALIK*
ADDRESS: *6560 Hollywood Blvd.*
CITY: *Woodridge, Illinois 60517*

The undersigned says that on or about *July 18, 2006*, at or about *2:00 p.m.* the Defendant did unlawfully commit the offense of **Failure to Provide Guards** in violation of *2006* **IPMC-306.1** as amended and adopted by reference in Section *8-1J-1(A)* of the *Village of Woodridge* Code; in that said Defendant, the owner of *6560 Hollywood Blvd., Woodridge, Illinois,* failed to provide guards on an interior* balcony** more than 30 inches above the floor or grade below in that *the balcony overlooking the family room is missing balusters.*

Chuck Schmidt
Complainant

Sworn to and Subscribed before Me
This *28th* Day of *August, 2006*

_____
Notary Public

 *Or, exterior.
**Or, stair, landing, porch, deck, ramp, or other walking surface.

IMPROPER HANDRAILS—IPMC 306.1

STATE OF *ILLINOIS*
COUNTY OF DUPAGE
VILLAGE OF WOODRIDGE
v.
NAME: *TIM HALIK*
ADDRESS: *6560 Hollywood Blvd.*
CITY: *Woodridge, Illinois 60517*

The undersigned says that on or about *October 3, 2006,* at or about *2:00 p.m.* the Defendant did unlawfully commit the offense of **Improper Handrail** in violation of ***2006* IPMC-306.1** as amended and adopted by reference in Section *8-1J-1(A)* of the *Village of Woodridge* Code; in that said Defendant, the owner of *6560 Hollywood Blvd., Woodridge, Illinois,* did install an improper handrail on an exterior* flight of stairs having more than four risers in that the handrail is *24 inches**** above the nosing of the tread.***

Chuck Schmidt
Complainant

Sworn to and Subscribed before Me
This *28th* Day of *October, 2006*

_____
Notary Public

  *Or, interior.

  **Insert measurement that violates the ordinance.

***Or, above the finished floor of the landing or walking surface.

SAMPLE COMPLAINT

IMPROPER GUARDS—IPMC 306.1

STATE OF *ILLINOIS*
COUNTY OF DUPAGE
VILLAGE OF WOODRIDGE
v.
NAME: *TIM HALIK*
ADDRESS: *6560 Hollywood Blvd.*
CITY: *Woodridge, Illinois 60517*

The undersigned says that on or about *October 3, 2006*, at or about *2:00 p.m.* the Defendant did unlawfully commit the offense of **Improper Guards** in violation of *2006* **IPMC-306.1** as amended and adopted by reference in Section *8-1J-1(A)* of the *Village of Woodridge* Code; in that said Defendant, the owner of *6560 Hollywood Blvd., Woodridge, Illinois,* did install an improper guard on an interior* balcony in that the guard is only *24 inches*** above the floor of the balcony.***

Chuck Schmidt
Complainant

Sworn to and Subscribed before Me
This *28th* Day of *October, 2006*

_____
Notary Public

 *Or, exterior.
 **Insert measurement that violates the ordinance.
***Or, stair, landing, porch, deck, ramp, or other walking surface.

IPMC 307   Rubbish and Garbage Violations

SAMPLE COMPLAINT

- -

ACCUMULATION OF RUBBISH OR GARBAGE—IPMC 307.1

STATE OF *ILLINOIS*
COUNTY OF DUPAGE
VILLAGE OF WOODRIDGE
v.
NAME: *SANDI MUELLER*
ADDRESS: *4110 Winthrop Blvd*
CITY: *Woodridge, Illinois 60517*

The undersigned says that on or about *November 18, 2006*, at or about *3:00 p.m.* the Defendant did unlawfully commit the offense of **Accumulation of Rubbish or Garbage** in violation of ***2006* IPMC-307.1** as amended and adopted by reference in Section *8-1J-1(A)* of the *Village of Woodridge* Code; in that said Defendant, the owner of *4110 Winthrop Blvd., Woodridge, Illinois,* failed to keep the interior* of the premises free from the accumulation of rubbish,** being combustible waste material,*** in the interior of the structure in that the *master bedroom and living room were filled with newspapers, cartons, boxes, and magazines.*

Stacey Crockatt
Complainant

Sworn to and Subscribed before Me
This *8th* Day of *December, 2006*

_____
Notary Public

  *Or, exterior.
 **Or, garbage.
***Or, being animal or vegetable waste.

COMMENT: Inspectors have a tendency to substitute the words junk or debris for rubbish or garbage. Because rubbish and garbage are strictly defined in the IPMC, only those words should be used. See Section VI in Chapter 2 to help determine whether the substance is rubbish or garbage.

IMPROPER DISPOSAL OF RUBBISH —IPMC 307.2

STATE OF *ILLINOIS*
COUNTY OF DUPAGE
VILLAGE OF WOODRIDGE
v.
NAME: *SANDI MUELLER*
ADDRESS: *4110 Winthrop Blvd*
CITY: *Woodridge, Illinois 60517*

The undersigned says that on or about *November 18, 2006*, at or about *3:00 p.m.* the Defendant did unlawfully commit the offense of **Improper Disposal of Rubbish** in violation of *2006* **IPMC-307.2** as amended and adopted by reference in Section *8-1J-1(A)* of the *Village of Woodridge* Code; in that said Defendant, being the occupant of *4110 Winthrop Blvd., Woodridge, Illinois,* failed to dispose of rubbish in a clean and sanitary manner in approved containers in that *bags of rubbish are scattered throughout the backyard.*

Stacey Crockatt

Complainant

Sworn to and Subscribed before Me
This *8th* Day of *December, 2006*

_____

Notary Public

SAMPLE COMPLAINT

FAILURE TO PROVIDE APPROVED COVERED CONTAINERS—IPMC 307.2.1

STATE OF *ILLINOIS*
COUNTY OF DUPAGE
VILLAGE OF WOODRIDGE
v.
NAME: *LARRY SALG*
ADDRESS: *3891 Schiller Dr.*
CITY: *Woodridge, Illinois 60517*

The undersigned says that on or about *May 24, 2006*, at or about *2:00 p.m.* the Defendant did unlawfully commit the offense of **Failure to Provide Approved Covered Containers** in violation of *2006* **IPMC-307.2.1** as amended and adopted by reference in Section *8-1J-1(A)* of the *Village of Woodridge* Code; in that said Defendant, the owner of *34 West Reber St., Woodridge, Illinois,* failed to provide approved covered containers for rubbish for the occupants at said premises.

Don Lay

Complainant

Sworn to and Subscribed before Me
This *17th* Day of *June, 2006*

_____

Notary Public

FAILURE TO REMOVE RUBBISH—IPMC 307.2.1

STATE OF *ILLINOIS*
COUNTY OF DUPAGE
VILLAGE OF WOODRIDGE
v.
NAME: *LARRY SALG*
ADDRESS: *3891 Schiller Dr.*
CITY: *Woodridge, Illinois 60517*

The undersigned says that on or about *May 24, 2006,* at or about *2:00 p.m.* the Defendant did unlawfully commit the offense of **Failure to Remove Rubbish** in violation of *2006* **IPMC-307.2.1** as amended and adopted by reference in Section *8-1J-1(A)* of the *Village of Woodridge* Code; in that said Defendant, the owner of *34 West Reber St., Woodridge, Illinois,* failed to remove rubbish at said premises in that *there has been no rubbish removal for four weeks.*

Don Lay
Complainant

Sworn to and Subscribed before Me
This *17th* Day of *June, 2006*

_____

Notary Public

FAILURE TO REMOVE DOOR FROM REFRIGERATOR—IPMC 307.2.2

STATE OF *ILLINOIS*
COUNTY OF DUPAGE
VILLAGE OF WOODRIDGE
v.
NAME: *SANDI MUELLER*
ADDRESS: *4110 Winthrop Blvd*
CITY: *Woodridge, Illinois 60517*

The undersigned says that on or about *November 18, 2006*, at or about *3:00 p.m.* the Defendant did unlawfully commit the offense of **Failure to Remove Door from Refrigerator*** in violation of ***2006* IPMC-307.2.2** as amended and adopted by reference in Section *8-1J-1(A)* of the *Village of Woodridge* Code; in that said Defendant, being the *occupant* of *4110 Winthrop Blvd., Woodridge, Illinois,* failed to remove the door from a refrigerator* not in operation before storing** it on the premises *on the front lawn.*

Stacey Crockatt
Complainant

Sworn to and Subscribed before Me
This *8th* Day of *December, 2006*

_____
Notary Public

  *Or, equipment similar to a refrigerator.
**Or, discarded or abandoned.

- -

IMPROPER DISPOSAL OF GARBAGE—IPMC 307.3

STATE OF *ILLINOIS*
COUNTY OF DUPAGE
VILLAGE OF WOODRIDGE
v.
NAME: *SANDI MUELLER*
ADDRESS: *4110 Winthrop Blvd*
CITY: *Woodridge, Illinois 60517*

The undersigned says that on or about *November 18, 2006*, at or about *3:00 p.m.* the Defendant did unlawfully commit the offense of **Improper Disposal of Garbage** in violation of *2006* **IPMC-307.3** as amended and adopted by reference in Section *8-1J-1(A)* of the *Village of Woodridge* Code; in that said Defendant, being the occupant of *4110 Winthrop Blvd., Woodridge, Illinois,* failed to dispose of garbage in a clean and sanitary manner in an approved garbage disposal facility or approved garbage container in that *garbage, being rotten food, was left in the corners of the kitchen.*

Stacey Crockatt

Complainant

Sworn to and Subscribed before Me
This *8th* Day of *December, 2006*

_____

Notary Public

FAILURE TO PROVIDE GARBAGE FACILITIES—IPMC 307.3.1

STATE OF *ILLINOIS*
COUNTY OF DUPAGE
VILLAGE OF WOODRIDGE
v.
NAME: *LARRY SALG*
ADDRESS: *3891 Schiller St.*
CITY: *Woodridge, Illinois 60517*

The undersigned says that on or about *May 24, 2006*, at or about *2:00 p.m.* the Defendant did unlawfully commit the offense of **Failure to Provide Garbage Facilities** in violation of *2006* **IPMC-307.3.1** as amended and adopted by reference in Section *8-1J-1(A)* of the *Village of Woodridge* Code; in that said Defendant, the owner of *34 West Reber St., Woodridge, Illinois,* failed to provide approved, leakproof, covered, outside garbage containers* for the premises in that *the garbage containers had no lids and were cracked.*

Don Lay
Complainant

Sworn to and Subscribed before Me
This *17th* Day of June, *2006*

_____

Notary Public

*Or, an approved mechanical food waste grinder in the dwelling unit, or an approved incinerator unit in the structure available to the occupants in each dwelling unit.

FAILURE TO USE CONTAINERS—IPMC 307.3.2

STATE OF *ILLINOIS*
COUNTY OF DUPAGE
VILLAGE OF WOODRIDGE
v.
NAME: *KYLE'S CHICKEN SHACK, INC.*
ADDRESS: *109 Fremont St.*
CITY: *Woodridge, Illinois 60517*

The undersigned says that on or about *April 2, 2006*, at or about *3:00 p.m.* the Defendant did unlawfully commit the offense of **Failure to Use Containers** in violation of *2006* **IPMC-307.3.2** as amended and adopted by reference in Section *8-1J-1(A)* of the *Village of Woodridge* Code, in that said Defendant, being the operator of *Kyle's Chicken Shack and Ribs Restaurant,* an establishment producing garbage, located at *109 Fremont St., Woodridge, Illinois,* failed to provide and utilize approved leakproof containers with close-fitting covers for the storage of garbage in that *bags of garbage from the restaurant overflowed the top of the dumpster at the back of the restaurant and spilled onto the ground.*

Joan Rogers
Complainant

Sworn to and Subscribed before Me
This *30th* Day of *April, 2006*

_____
Notary Public

IPMC 308   Extermination Violations

IPMC 308.1—Infestation

- -

FAILURE TO KEEP STRUCTURE FREE FROM INFESTATION—IPMC 308.1

STATE OF *ILLINOIS*
COUNTY OF DUPAGE
VILLAGE OF WOODRIDGE
v.
NAME: *KYLE'S CHICKEN SHACK, INC.*
ADDRESS: *109 Fremont St.*
CITY: *Woodridge, Illinois 60517*

The undersigned says that on or about *April 2, 2006*, at or about *3:00 p.m.* the Defendant did unlawfully commit the offense of **Failure to Keep Structure Free from Insect* Infestation** in violation of **2006 IPMC-308.1** as amended and adopted by reference in Section *8-1J-1(A)* of the *Village of Woodridge* Code, in that said Defendant, the owner of *109 Fremont St., Woodridge, Illinois*, failed to keep the structure free from insect* infestation in that *there are roaches in the kitchen and bathroom areas of the restaurant.*

Joan Rogers

Complainant

Sworn to and Subscribed before Me
This *30th* Day of *April, 2006*

_____

Notary Public

*Or, rodent.

SAMPLE COMPLAINT

FAILURE TO EXTERMINATE STRUCTURE AFTER INFESTATION—IPMC 308.1

STATE OF *ILLINOIS*
COUNTY OF DUPAGE
VILLAGE OF WOODRIDGE
v.
NAME: *KYLE'S CHICKEN SHACK, INC.*
ADDRESS: *109 Fremont St.*
CITY: *Woodridge, Illinois 60517*

The undersigned says that on or about *May 2, 2006*, at or about *3:00 p.m.* the Defendant did unlawfully commit the offense of **Failure to Exterminate Structure After Insect* Infestation** in violation of *2006* **IPMC-308.1** as amended and adopted by reference in Section *8-1J-1(A)* of the *Village of Woodridge* Code, in that said Defendant, the owner of *109 Fremont St., Woodridge, Illinois,* failed to exterminate the structure after insect* infestation was found in *the kitchen and bathroom areas of the restaurant.*

Joan Rogers
Complainant

Sworn to and Subscribed before Me
This *30th* Day of *May, 2006*

_____
Notary Public

*Or, rodent.

- -

FAILURE TO TAKE PRECAUTIONS TO PREVENT REINFESTATION—IPMC 308.1

STATE OF *ILLINOIS*
COUNTY OF DUPAGE
VILLAGE OF WOODRIDGE
v.
NAME: *KYLE'S CHICKEN SHACK, INC.*
ADDRESS: *109 Fremont St.*
CITY: *Woodridge, Illinois 60517*

The undersigned says that on or about *May 2, 2006*, at or about *3:00 p.m.* the Defendant did unlawfully commit the offense of **Failure to Take Precautions to Prevent Reinfestation** in violation of ***2006* IPMC-308.1** as amended and adopted by reference in Section *8-1J-1(A)* of the *Village of Woodridge* Code, in that said Defendant, the owner of *109 Fremont St., Woodridge, Illinois*, failed to take precautions to prevent the reinfestation of insects* of the structure after extermination in that *there is open garbage and food products in the kitchen area of the restaurant.*

Joan Rogers
Complainant

Sworn to and Subscribed before Me
This *30th* Day of *May, 2006*

_____
Notary Public

*Or, rodents.

TABLE 3-1

WHO IS RESPONSIBLE FOR THE EXTERMINATION OF INSECTS AND RODENTS IN A STRUCTURE?

	Owner	Occupant
Prior to renting or leasing the structure—IPMC 308.2	X	
One-family dwelling—IPMC 308.3		X
Single-tenant nonresidential structure—IPMC 308.3		X
Public or shared areas of structure containing two or more dwelling units and the exterior—IPMC 308.4	X	
Public or shared areas of a multiple occupancy structure and the exterior—IPMC 308.4	X	
Public or shared areas of a rooming house and the exterior—IPMC 308.4	X	
Public or shared areas of a nonresidential structure and the exterior—IPMC 308.4	X	
Infestation caused by failure of occupant to prevent infestation in the area occupied—IPMC 308.4		X
Maintaining structure in rodent and pest-free condition—IPMC 308.5		X
Infestation caused by defects in structure—IPMC 308.5	X	

Light, Ventilation, and Occupancy Limits

CHAPTER 4 covers violations in existing structures that fail to meet the minimum standards for light, ventilation, and space. The chapter prescribes minimum standards for light in habitable spaces, common halls and stairways, and other spaces. It also regulates the type of ventilation needed for habitable spaces, bathrooms, cooking facilities, processing space, and clothes dryers. The chapter sets limits on the number of people that can occupy dwellings, and sets minimum standards for the width and height of rooms.

IPMC 401   General Violations

IPMC 401.2—Responsibility

SAMPLE COMPLAINT

- -

OCCUPYING PREMISES IN VIOLATION OF LIGHT* CONDITIONS—IPMC 401.2

STATE OF *ILLINOIS*
COUNTY OF DUPAGE
VILLAGE OF WOODRIDGE
v.
NAME: *ROB MC GINNIS*
ADDRESS: *801 Saratoga St.*
CITY: *Woodridge, Illinois 60517*

The undersigned says that on or about *November 18, 2006*, at or about *3:00 p.m.* the Defendant, *Rob McGinnis,* did unlawfully commit the offense **Occupying Premises in Violation of Light* Conditions** in violation of *2006* **IPMC-401.2** as amended and adopted by reference in Section *8-1J-1(A)* of the *Village of Woodridge* Code; in that said Defendant, the owner-occupant, occupied the premises of *801 Saratoga St., Unit 1A, Woodridge, Illinois,* at a time when the premises were in violation of *2006 IPMC-402.2* as amended and adopted by reference in Section *8-1J-1(A)* of the *Village of Woodridge* Code in that *the common staircase from the first floor to the second floor has a broken light fixture and so there is no illumination in that area.*

Kathy Hejnicki

Complainant

Sworn to and Subscribed before Me
This *8th* Day of *December, 2006*

_____

Notary Public

*Or, ventilation or space.

ALLOWING PERSONS TO OCCUPY PREMISES IN VIOLATION OF LIGHT* CONDITIONS—IPMC 401.2

STATE OF *ILLINOIS*
COUNTY OF DUPAGE
VILLAGE OF WOODRIDGE
v.
NAME: *ROB MC GINNIS*
ADDRESS: *801 Saratoga St.*
CITY: *Woodridge, Illinois 60517*

The undersigned says that on or about *November 18, 2006,* at or about *3:00 p.m.* the Defendant, *Rob McGinnis,* did unlawfully commit the offense **Allowing the Occupation of a Premises in Violation of Light* Conditions** in violation of *2006* **IPMC-401.2** as amended and adopted by reference in Section *8-1J-1(A)* of the *Village of Woodridge* Code; in that said Defendant, the owner, allowed *John Fincham* to occupy the premises of *801 Saratoga St., Unit 2A, Woodridge, Illinois,* at a time when the premises were in violation of *2006 IPMC-402.2* as amended and adopted by reference in Section *8-1J-1(A)* of the *Village of Woodridge* Code in that *the common staircase from the first floor to the second floor has a broken light fixture and so there is no illumination in that area.*

Kathy Hejnicki

Complainant

Sworn to and Subscribed before Me
This *8th* Day of *December, 2006*

_____

Notary Public

*Or, ventilation or space.

IPMC 402   Light Violations

TABLE 4-1

- -

LIGHT REQUIREMENTS—IPMC 402

Space	Requirement	Section
Every habitable space*	Must have at least one window of approved size facing directly to outdoors or to a court	402.1
Minimum total glazed space for every habitable space	8 percent of the floor area of the room**	402.1
Rooms or spaces without exterior glazing where natural light provided by adjoining room	Unobstructed opening of adjoining room must be 8 percent of floor area of the interior room or space but not less than 25 square feet (2.33m²)	402.1 Exception
Exterior glazing area	Based on total floor area being served	
Common hall and stairway for residential occupancy other than one and two family dwelling	Must be lighted at all times with at least 60-watt standard incandescent light bulb for each 200 square feet(19 m²) of floor area or equivalent and spaced not greater than 30 feet (9144 mm) apart	402.2
In non-residential occupancies, means of egress, including exterior means of egress stairways	Must be lighted at all times with 1 footcandle (11 lux) at floors, landings and treads	402.2
All other spaces	Sufficient light to permit maintenance of sanitary conditions and safe occupancy of space and utilization of appliances, equipment and fixtures	402.3

*Habitable space is any space in a structure for living, sleeping, eating, or cooking but not a bathroom, toilet room, closet, hall, storage or utility space, or similar area, see IPMC Chapter 2, Section 202, Definitions.

**If a wall or portion of a structure face a window of any room and are less than 3 feet (914 mm) from the window and extend to a level above that of the ceiling of the room, the window is not deemed to face directly to the outdoors nor a court, and cannot be included as contributing to the required minimum total window area for the room.

SAMPLE COMPLAINT

FAILURE TO PROVIDE LIGHT IN HABITABLE SPACE—IPMC 402.1

STATE OF *ILLINOIS*
COUNTY OF DUPAGE
VILLAGE OF WOODRIDGE
v.
NAME: *ROB MC GINNIS*
ADDRESS: *801 Saratoga St.*
CITY: *Woodridge, Illinois 60517*

The undersigned says that on or about *November 18, 2006*, at or about *3:00 p.m.* the Defendant did unlawfully commit the offense of **Failure to Provide Light in Habitable Space** in violation of *2006* **IPMC-402.1** as amended and adopted by reference in Section *8-1J-1(A)* of the *Village of Woodridge* Code, in that said Defendant, the owner of *801 Saratoga St., Unit 3A, Woodridge, Illinois*, failed to provide sufficient light *for a bedroom of 100 square feet* in the structure in that *the window provided is only four square feet*.

Kathy Hejnicki

Complainant

Sworn to and Subscribed before Me
This *8th* Day of *December, 2006*

_____

Notary Public

FAILURE TO PROVIDE LIGHT IN COMMON HALL*—IPMC 402.2

STATE OF *ILLINOIS*
COUNTY OF DUPAGE
VILLAGE OF WOODRIDGE
v.
NAME: *ROB MC GINNIS*
ADDRESS: *801 Saratoga St.*
CITY: *Woodridge, Illinois 60517*

The undersigned says that on or about *November 18, 2006*, at or about *3:00 p.m.* the Defendant did unlawfully commit the offense of **Failure to Provide Light in Common Hall*** in violation of ***2006* IPMC-402.2** as amended and adopted by reference in Section *8-1J-1(A)* of the *Village of Woodridge* Code, in that said Defendant, the owner of *801 Saratoga St., Woodridge, Illinois*, failed to provide sufficient light in the premises, a multi-family residence, in that *the common hallway on the first floor has a broken light fixture and so there is no illumination in that area.*

Kathy Hejnicki

Complainant

Sworn to and Subscribed before Me
This *8th* Day of *December, 2006*

_____

Notary Public

*Or, stairways.

FAILURE TO PROVIDE LIGHT FOR MEANS OF EGRESS—IPMC 402.2

STATE OF *ILLINOIS*
COUNTY OF DUPAGE
VILLAGE OF WOODRIDGE
v.
NAME: *KYLE'S CHICKEN SHACK, INC.*
ADDRESS: *109 Fremont St.*
CITY: *Woodridge, Illinois 60517*

The undersigned says that on or about *November 18, 2006*, at or about *3:00 p.m.* the Defendant did unlawfully commit the offense of **Failure to Provide Light for Means of Egress** in violation of *2006* **IPMC-402.2** as amended and adopted by reference in Section *8-1J-1(A)* of the *Village of Woodridge* Code, in that said Defendant, the owner of *109 Fremont St., Woodridge, Illinois,* failed to provide sufficient light for the *rear exterior stairway*, in that *there is no light for any part of the staircase.*

Kathy Hejnicki
Complainant

Sworn to and Subscribed before Me
This *8th* Day of *December, 2006*

_____
Notary Public

SAMPLE COMPLAINT

FAILURE TO PROVIDE LIGHT—IPMC 402.3

STATE OF *ILLINOIS*
COUNTY OF DUPAGE
VILLAGE OF WOODRIDGE
v.
NAME: *ROB MC GINNIS*
ADDRESS: *801 Saratoga St.*
CITY: *Woodridge, Illinois 60517*

The undersigned says that on or about *November 18, 2006*, at or about *3:00 p.m.* the Defendant did unlawfully commit the offense of **Failure to Provide Light** in violation of *2006* **IPMC-402.3** as amended and adopted by reference in Section *8-1J-1(A)* of the *Village of Woodridge* Code, in that said Defendant, the owner of *801 Saratoga St., Woodridge, Illinois*, failed to provide sufficient light for the *bathroom in Unit 2A*, in that *the light fixture is broken*, thereby interfering with the ability to maintain the *bathroom* in a sanitary condition, and the safe occupancy of the space and utilization of the appliances, equipment, or fixtures.

Kathy Hejnicki

Complainant

Sworn to and Subscribed before Me
This *8th* Day of *December, 2006*

_____

Notary Public

IPMC 403　Ventilation Violations

TABLE 4-2

VENTILATION REQUIREMENTS—IPMC 403

Space or Object	Requirement	Section
Every habitable space*	Must have at least one openable window	403.1
Openable area of window in every room	Must equal 45 percent of required minimum glazed area (8 percent of the floor area of the room)	403.1
Rooms or spaces without openings to the outdoors ventilated by adjoining room	Unobstructed opening of adjoining room must be 8 percent of floor area of the interior room or space but not less than 25 square feet (2.33 m²)	403.1 Exception
Ventilation openings to the outdoors	Based on a total floor area being ventilated	
Bathrooms and toilet rooms	Must have at least one openable window unless equipped with mechanical ventilation system vented to the outside and not recirculated	403.2
Cooking or cooking facilities**	Not allowed in rooming or dormitory unit unless allowed by certificate of occupancy or approved in writing by code official	403.3
Areas generating injurious, toxic, irritating or noxious fumes, gases, dusts, or mists	Local exhaust system vented to the exterior and not recirculated	403.4
Clothes dryers	Independent system exhausted according to manufacturer's instructions	403.5

*Habitable space is any space in a structure for living, sleeping, eating, or cooking but not a bathroom, toilet room, closet, hall, storage or utility space, or similar area. See IPMC Chapter 2, Section 202, Definitions.

**Does not include coffee pots and microwave ovens as cooking appliances in the 2006 IPMC.

FAILURE TO PROVIDE VENTILATION IN HABITABLE SPACE—IPMC 403.1

STATE OF *ILLINOIS*
COUNTY OF DUPAGE
VILLAGE OF WOODRIDGE
v.
NAME: *ROB MC GINNIS*
ADDRESS: *801 Saratoga St.*
CITY: *Woodridge, Illinois 60517*

The undersigned says that on or about *November 18, 2006,* at or about *3:00 p.m.* the Defendant did unlawfully commit the offense of **Failure to Provide Ventilation in Habitable Space** in violation of *2006* **IPMC-403.1** as amended and adopted by reference in Section *8-1J-1(A)* of the *Village of Woodridge* Code, in that said Defendant, the owner of *801 Saratoga St., Woodridge, Illinois,* failed to provide an openable window in the *dining room in Unit 2A,* in that *window is painted shut.*

Kathy Hejnicki

Complainant

Sworn to and Subscribed before Me
This *8th* Day of *December, 2006*

_____

Notary Public

FAILURE TO PROVIDE VENTILATION IN BATHROOM*—IPMC 403.2

STATE OF *ILLINOIS*
COUNTY OF DUPAGE
VILLAGE OF WOODRIDGE
v.
NAME: *ROB MC GINNIS*
ADDRESS: *801 Saratoga St.*
CITY: *Woodridge, Illinois 60517*

The undersigned says that on or about *November 18, 2006*, at or about *3:00 p.m.* the Defendant did unlawfully commit the offense of **Failure to Provide Ventilation in Bathroom*** in violation of *2006* **IPMC-403.2** as amended and adopted by reference in Section *8-1J-1(A)* of the *Village of Woodridge* Code, in that said Defendant, the owner of *801 Saratoga St., Woodridge, Illinois,* failed to provide proper ventilation in the *bathroom in Unit 2A,* in that *the mechanical venting system does not discharge air to the outdoors.*

Kathy Hejnicki
Complainant

Sworn to and Subscribed before Me
This *8th* Day of *December, 2006*

_____

Notary Public

*Or, toilet room.

SAMPLE COMPLAINT

COOKING IN A ROOMING UNIT*—IPMC 403.3

STATE OF *ILLINOIS*
COUNTY OF DUPAGE
VILLAGE OF WOODRIDGE
v.
NAME: *JOHN FINCHAM*
ADDRESS: *801 Saratoga St.*
CITY: *Woodridge, Illinois 60517*

The undersigned says that on or about *November 18, 2006*, at or about *3:00 p.m.* the Defendant did unlawfully commit the offense of **Cooking in a Rooming Unit*** in violation of ***2006 IPMC-403.3*** as amended and adopted by reference in Section *8-1J-1(A)* of the *Village of Woodridge* Code, in that said Defendant, the occupant of *801 Saratoga St., Woodridge, Illinois, Unit 2A,* a rooming unit, used a cooking appliance, *being an electric grill,* for cooking, within said unit.

Kathy Hejnicki

Complainant

Sworn to and Subscribed before Me
This *8th* Day of *December, 2006*

_____

Notary Public

*Or, dormitory unit.

COMMENT: The code official may grant an exception in writing for cooking in a rooming unit or dormitory unit and, under the 2006 IPMC. Devices such as coffee pots and microwave ovens are not considered cooking appliances.

FAILURE TO PROVIDE PROCESS VENTILATION—IPMC 403.4

STATE OF *ILLINOIS*
COUNTY OF DUPAGE
VILLAGE OF WOODRIDGE
v.
NAME: *KYLE'S CHICKEN SHACK, INC.*
ADDRESS: *109 Fremont St.*
CITY: *Woodridge, Illinois 60517*

The undersigned says that on or about *October 3, 2006*, at or about *3:00 p.m.* the Defendant did unlawfully commit the offense of **Failure to Provide Process Ventilation** in violation of **2006 IPMC-403.4** as amended and adopted by reference in Section *8-1J-1(A)* of the *Village of Woodridge* Code, in that said Defendant, the owner of *109 Fremont St., Woodridge, Illinois*, failed to provide proper process ventilation for gases* being generated, in that *the local exhaust ventilation system recirculates into the attic area.*

Kathy Hernicki
_____
Complainant

Sworn to and Subscribed before Me
This *28th* Day of *October, 2006*

_____
Notary Public

*Or, injurious, toxic, irritating or noxious fumes, dust or mists.

SAMPLE COMPLAINT

IMPROPER CLOTHES DRYER EXHAUST—IPMC 403.5

STATE OF *ILLINOIS*
COUNTY OF DUPAGE
VILLAGE OF WOODRIDGE
v.
NAME: *ROB MC GINNIS*
ADDRESS: *801 Saratoga St.*
CITY: *Woodridge, Illinois 60517*

The undersigned says that on or about *November 18, 2006,* at or about *3:00 p.m.* the Defendant did unlawfully commit the offense of **Improper Clothes Dryer Exhaust** in violation of *2006* **IPMC-403.5** as amended and adopted by reference in Section *8-1J-1(A)* of the *Village of Woodridge* Code, in that said Defendant, the owner of *801 Saratoga St., Woodridge, Illinois,* failed to provide proper ventilation for the clothes dryer *in the basement* in that *the exhaust system is not vented to the outdoors pursuant to the manufacturer's specifications.*

Kathy Hejnicki
Complainant

Sworn to and Subscribed before Me
This *8th* Day of *December, 2006*

_____

Notary Public

*Or, toilet room.

IPMC 404   Occupancy Limitations Violations

FAILURE TO PROVIDE PRIVATE SPACE—IPMC 404.1

STATE OF *ILLINOIS*
COUNTY OF DUPAGE
VILLAGE OF WOODRIDGE
v.
NAME: *LARRY SALG*
ADDRESS: *3859 Schiller Dr.*
CITY: *Woodridge, Illinois 60517*

The undersigned says that on or about *May 24, 2006,* at or about *9:00 a.m.* the Defendant did unlawfully commit the offense of **Failure to Provide Private Space** in violation of ***2006 IPMC-404.1*** as amended and adopted by reference in Section *8-1J-1(A)* of the *Village of Woodridge* Code, in that said Defendant, the owner of *34 West Reber St., Woodridge, Illinois,* failed to provide privacy for a dwelling unit,* being *Apartment 7G,* in that *there is no door on the unit separating it from the adjoining hallway.*

Karyn Byrne
Complainant

Sworn to and Subscribed before Me
This *17th* Day of *June, 2006*

_____
Notary Public

*Or, hotel unit, housekeeping unit, rooming unit, or dormitory unit.

TABLE 4-2

OCCUPANCY LIMITATIONS: ROOM MEASUREMENTS—2003 IPMC

Room	Measurement	Code Section
Habitable room width other than kitchen	Not less than 7 feet (2134 mm) in any plan dimension	404.2
Kitchen passageway	Not less than 3 feet (914 mm) between counterfronts and appliances or counterfronts and walls	404.2
Ceiling height in habitable spaces, hallways, corridors, laundry areas, bathrooms, toilet rooms and habitable basement areas	Not less than 7 feet (2134 mm)	404.3
Ceiling height in one- and two-family dwellings	Beams or girders spaced not less than 4 feet (1219 mm) on center and projecting not more than 6 inches (152 mm) below the required ceiling height	404.3 Exception 1
Ceiling height in basement rooms in one- and two-family dwellings used only for laundry, study., or recreation	Not less than 6 feet 8 inches (2033 mm) with not less than 6 feet 4 inches (1932 mm) of clear height under beams, girders, ducts, and similar obstructions	404.3 Exception 2
Ceiling height in rooms occupied exclusively for sleeping, study, or similar purposes and having a sloped ceiling over all or part of the room	At least 7 feet (2134 mm) over not less than one-third of the required minimum floor area*	404.3 Exception 3
Bedroom—1 person occupancy	70 square feet (6.5 m²) of floor area	404.4.1
Bedroom—More than 1 person occupancy	50 square feet (4.6 m²) of floor area for each occupant	404.4.1
Efficiency living unit by not more than two occupants	Clear floor area of not less than 220 square feet (20.4 m²) exclusive of kitchen and bath areas	404.6(1)
Efficiency living unit by three occupants	Clear floor area of not less than 320 square feet (29.7 m²) exclusive of kitchen and bath areas	404.6(2)
Efficiency living units	Clear working space of not less than 30 inches (762 mm) in front for required kitchen sink, cooking appliance, and refrigeration facilities	404.6(2)

*Floor area: Only those portions of the floor area with a clear ceiling height of 5 feet (1524 mm) or more are included.

TABLE 4-3

OCCUPANCY LIMITATIONS: ROOM MEASUREMENTS—2006 IPMC

Room	Measurement	Code Section
Habitable room width other than kitchen	Not less than 7 feet (2134 mm) in any plan dimension	404.2
Kitchen passageway	Not less than 3 feet (914 mm) between counterfronts and appliances or counterfronts and walls	404.2
Ceiling height in habitable spaces, hallways, corridors, laundry areas, bathrooms, toilet rooms, and habitable basement areas	Not less than 7 feet (2134 mm)	404.3
Ceiling height in one- and two-family dwellings	Beams or girders spaced not less than 4 feet (1219 mm) on center and projecting not more than 6 inches (152 mm) below the required ceiling height	404.3 Exception 1
Ceiling height in basement rooms in one- and two-family dwellings used only for laundry, study, or recreation	Not less than 6 feet 8 inches (2033 mm) with not less than 6 feet 4 inches (1932 mm) of clear height under beams, girders, ducts and similar obstructions	404.3 Exception 2
Ceiling height in rooms occupied exclusively for sleeping, study, or similar purposes and having a sloped ceiling over all or part of the room	At least 7 feet (2134 mm) over not less than one-third of the required minimum floor area*	404.3 Exception 3
Bedroom	70 square feet (6.5 m²) of floor area	404.4.1
Living room	120 square feet (11.2 m²)	404.4.1
Efficiency living unit by not more than two occupants	Clear floor area of not less than 220 square feet (20.4 m²) exclusive of kitchen and bath areas	404.6(1)
Efficiency living unit by three occupants	Clear floor area of not less than 320 square feet (29.7 m²) exclusive of kitchen and bath areas	404.6(2)
Efficiency living units	Clear working space of not less than 30 inches (762 mm) in front for required kitchen sink, cooking appliance and refrigeration facilities	404.6(2)

*Floor area: Only those portions of the floor area with a clear ceiling height of 5 feet (1524 mm) or more are included.

TABLE 4-4

OCCUPANCY LIMITATIONS—ACCESS REQUIREMENTS

Area	Requirement	Code Section
Kitchen	Passageway not less than 3 feet (914 mm) between counterfronts and appliances or counterfronts and walls	404.2
Bedroom	May not be only means of access to other bedrooms or habitable spaces unless unit has fewer than two bedrooms	404.4.2
Water closet	Each bedroom must have access to one water closet and one lavatory located on same story as bedroom or an adjacent story without passing through another bedroom	404.4.3

TABLE 4-5

OCCUPANCY LIMITATIONS—SLEEPING RULES—2003 IPMC

Area	Sleeping Allowed		Code Section
	Yes	No	
Bedroom—proper size.	X		404.4
Bathroom		X	404.4.4
Kitchen		X	404.4.4
Living room*—1–2 occupants	X		404.5
Living room under 120 square feet—3–5 occupants		X	404.5
Living room* with at least 120 square feet—3–5 occupants	X**		404.5
Living room with at less than 150 square feet—6 or more occupants		X	404.5
Living room* with at least 150 square feet—6 or more occupants	X**		404.5
Dining room*—1–2 occupants	X**		404.5
Dining room under 80 square feet—3–5 occupants		X	404.5
Dining room* with at least 80 square feet—3–5 occupants	X**		404.5
Dining room with less than 100 square feet—6 or more occupants		X	404.5
Dining room* with at least 100 square feet—6 or more occupants	X**		404.5
Living/dining room*—1–2 occupants	X**		404.5
Living/dining room under 200 square feet—3–5 occupants		X	404.5
Living/dining room* with at least 200 square feet—3–5 occupants	X**		404.5
Living/dining room with less than 250 square feet—6 or more occupants		X	404.5
Living/dining room* with at least 250 square feet—6 or more occupants	X**		404.5

*This assumes that the room is not the only means of access to other bedrooms or habitable spaces and does not serve as the only means of egress from other habitable spaces except for units containing fewer than two bedrooms, and that it meets the minimum square footage required for sleeping in 404.4.1. If not, it may not be used for sleeping. (See 404.4.2)

**To use a living and/or dining room area for sleeping purposes, the additional sleeping area space must be calculated separately from the minimum area required for living and dining rooms. For example, in a living/dining room suitable for 3 to 5 occupants (200 square feet), 270 square feet is necessary for one person to sleep in that room, 300 square feet for two persons and 350 for three persons, 400 for four persons, 450 for five persons, and so forth.

TABLE 4-6

OCCUPANCY LIMITATIONS—SLEEPING RULES—2006 IPMC

Area	Sleeping Allowed		Code Section
	Yes	No	
Bedroom*—at least 70 square feet (6.5 m²)	X		404.4.41
Bathroom		X	404.4.4
Kitchen		X	404.4.4
Other uninhabitable spaces		X	404.4.4
Living room—at least 120 square feet (11.2 m²)	X		404.4.1

*This assumes that the room is not the only means of access to other bedrooms or habitable spaces and does not serve as the only means of egress from other habitable spaces except for units containing fewer than two bedrooms, and that it meets the minimum square footage required for sleeping in 404.4.1. If not, it may not be used for sleeping. (See 404.4.2)

SAMPLE COMPLAINT

VIOLATION OF MINIMUM ROOM WIDTH REQUIREMENTS—IPMC 404.2

STATE OF *ILLINOIS*
COUNTY OF DUPAGE
VILLAGE OF WOODRIDGE
v.
NAME: *LARRY SALG*
ADDRESS: *3859 Schiller Dr.*
CITY: *Woodridge, Illinois 60517*

The undersigned says that on or about *May 24, 2006,* at or about *9:00 a.m.* the Defendant did unlawfully commit the offense of **Violation of Minimum Room Width Requirements** in violation of ***2006* IPMC-404.2** as amended and adopted by reference in Section *8-1J-1(A)* of the *Village of Woodridge* Code, in that said Defendant, the owner of *34 West Reber St., Woodridge, Illinois,* violated the minimum room width requirements for a habitable room, *being the bedroom in Apartment 7G,* in that the room width is only *6* feet, which is less than the 7 feet required.

Karyn Byrne

Complainant

Sworn to and Subscribed before Me
This *17th* Day of *June, 2006*

_____

Notary Public

VIOLATION OF MINIMUM ROOM CEILING HEIGHT REQUIREMENTS—IPMC 404.3

STATE OF *ILLINOIS*
COUNTY OF DUPAGE
VILLAGE OF WOODRIDGE
v.
NAME: *LARRY SALG*
ADDRESS: *3859 Schiller Dr.*
CITY: *Woodridge, Illinois 60517*

The undersigned says that on or about *May 24, 2006*, at or about *9:00 a.m.* the Defendant did unlawfully commit the offense of **Violation of Minimum Room Ceiling Height Requirements** in violation of *2006* **IPMC-404.3** as amended and adopted by reference in Section *8-1J-1(A)* of the *Village of Woodridge* Code, in that said Defendant, the owner of *34 West Reber St., Woodridge, Illinois*, violated the minimum room ceiling height requirements for habitable spaces, *being the laundry area, and dwelling units in the basement of the building,* in that the ceiling height is only *5.8* feet, which is less than the 7 feet required.

Karyn Byrne

Complainant

Sworn to and Subscribed before Me
This *17th* Day of *June, 2006*

_____

Notary Public

VIOLATION OF REQUIREMENTS FOR BEDROOMS (SQUARE FOOTAGE)—2003 IPMC 404.4.1

STATE OF *ILLINOIS*
COUNTY OF DUPAGE
VILLAGE OF WOODRIDGE
v.
NAME: *LARRY SALG*
ADDRESS: *3859 Schiller Dr.*
CITY: *Woodridge, Illinois 60517*

The undersigned says that on or about *May 24, 2006*, at or about *9:00 a.m.* the Defendant did unlawfully commit the offense of **Violation of Requirements for Bedrooms** in violation of **2003 IPMC-404.4.1** as amended and adopted by reference in Section *8-1J-1(A)* of the *Village of Woodridge* Code, in that said Defendant, the owner of *34 West Reber St., Woodridge, Illinois*, violated the space requirements for bedrooms, in *Apartment 7G*, in that the *second* bedroom only contains *42* square feet.

Karyn Byrne

Complainant

Sworn to and Subscribed before Me
This *17th* Day of *June, 2006*

_____

Notary Public

COMMENT: The 2003 IPMC requires 70 square feet (6.5 m^2) for the first person and 50 square feet (6.5 m^2) for each additional occupant for a bedroom.

SAMPLE COMPLAINT

- -

VIOLATION OF REQUIREMENTS FOR BEDROOMS (SQUARE FOOTAGE)—2006 IPMC 404.4.1

STATE OF *ILLINOIS*
COUNTY OF DUPAGE
VILLAGE OF WOODRIDGE
v.
NAME: *LARRY SALG*
ADDRESS: *3859 Schiller Dr.*
CITY: *Woodridge, Illinois 60517*

The undersigned says that on or about *May 24, 2006,* at or about *9:00 a.m.* the Defendant did unlawfully commit the offense of **Violation of Requirements for Bedrooms** in violation of **2006 IPMC-404.4.1** as amended and adopted by reference in Section *8-1J-1(A)* of the *Village of Woodridge* Code, in that said Defendant, the owner of *34 West Reber St., Woodridge, Illinois,* violated the space requirements for bedrooms, in *Apartment 7G,* in that the *second* bedroom only contains *42* square feet.

Karyn Byrne

Complainant

Sworn to and Subscribed before Me
This *17th* Day of *June, 2006*

_____

Notary Public

COMMENT: The 2006 IPMC requires 70 square feet (6.5 m²) for a bedroom without regard to the number of occupants.

VIOLATION OF REQUIREMENTS FOR BEDROOMS (NUMBER OF PEOPLE)—2003 IPMC 404.4.1

STATE OF *ILLINOIS*
COUNTY OF DUPAGE
VILLAGE OF WOODRIDGE
v.
NAME: *LARRY SALG*
ADDRESS: *3859 Schiller Dr.*
CITY: *Woodridge, Illinois 60517*

The undersigned says that on or about *May 24, 2006,* at or about *9:00 a.m.* the Defendant did unlawfully commit the offense of **Violation of Requirements for Bedrooms** in violation of **2003 IPMC-404.4.1** as amended and adopted by reference in Section *8-1J-1(A)* of the *Village of Woodridge* Code, in that said Defendant, the owner of *34 West Reber St., Woodridge, Illinois,* violated the space requirements bedrooms, in *Apartment 7G,* in that *three* people are sleeping in a bedroom that only contains *100* square feet.

Karyn Byrne
Complainant

Sworn to and Subscribed before Me
This *17th* Day of *June, 2006*

_____
Notary Public

COMMENT: The 2003 IPMC requires 70 square feet (6.5 m²) for the first person and 50 square feet (6.5 m²) for each additional occupant for a bedroom.

SAMPLE COMPLAINT

VIOLATION OF REQUIREMENTS FOR BEDROOMS (ACCESS)—IPMC 404.4.2

STATE OF *ILLINOIS*
COUNTY OF DUPAGE
VILLAGE OF WOODRIDGE
v.
NAME: *LARRY SALG*
ADDRESS: *3859 Schiller Dr.*
CITY: *Woodridge, Illinois 60517*

The undersigned says that on or about *May 24, 2006,* at or about *9:00 a.m.* the Defendant did unlawfully commit the offense of **Violation of Requirements for Bedrooms** in violation of **2006** **IPMC-404.4.2** as amended and adopted by reference in Section *8-1J-1(A)* of the *Village of Woodridge* Code, in that said Defendant, the owner of *34 West Reber St., Woodridge, Illinois,* violated the access from bedrooms requirement, in *Apartment 6B,* in that *one of the bedrooms* is the only means of access *to the second bedroom* and is the only means of egress *for said bedroom.*

Karyn Byrne

Complainant

Sworn to and Subscribed before Me
This *17th* Day of *June, 2006*

_____

Notary Public

VIOLATION OF REQUIREMENTS FOR WATER CLOSET—IPMC 404.4.3

STATE OF *ILLINOIS*
COUNTY OF DUPAGE
VILLAGE OF WOODRIDGE
v.
NAME: *LARRY SALG*
ADDRESS: *3859 Schiller Dr.*
CITY: *Woodridge, Illinois 60517*

The undersigned says that on or about *May 24, 2006,* at or about *9:00 a.m.* the Defendant did unlawfully commit the offense of **Violation of Requirements for Water Closets** in violation of *2006* **IPMC-404.4.3** as amended and adopted by reference in Section *8-1J-1(A)* of the *Village of Woodridge* Code, in that said Defendant, the owner of *34 West Reber St., Woodridge, Illinois*, violated the water closet access requirement, in *Apartment 6B*, in that in order to have access to a water closet and lavatory* from a bedroom, a person must pass through another bedroom.

Karyn Byrne

Complainant

Sworn to and Subscribed before Me
This *17th* Day of *June, 2006*

_____

Notary Public

*Or, must use the water closet and lavatory located more than one story away from the apartment.

SAMPLE COMPLAINT

SLEEPING IN A NONHABITABLE SPACE—IPMC 404.4.4

STATE OF *ILLINOIS*
COUNTY OF DUPAGE
VILLAGE OF WOODRIDGE
v.
NAME: *JOHN FINCHAM*
ADDRESS: *34 West Reber St. Apt. 4C*
CITY: *Woodridge, Illinois 60517*

The undersigned says that on or about *May 24, 2006,* at or about *9:00 a.m.* the Defendant did unlawfully commit the offense of **Sleeping in a Nonhabitable Space** in violation of **2006 IPMC-404.4.4** as amended and adopted by reference in Section *8-1J-1(A)* of the *Village of Woodridge* Code, in that said Defendant, the occupant of *34 West Reber St., Woodridge, Illinois, Apartment 4C,* used a nonhabitable space, being a *closet,** for sleeping purposes.

Karyn Byrne

Complainant

Sworn to and Subscribed before Me
This *17th* Day of *June, 2006*

_____

Notary Public

*Or, kitchen or other nonhabitable space (describe in detail).

OVERCROWDING—2003 IPMC 404.5

STATE OF *ILLINOIS*
COUNTY OF DUPAGE
VILLAGE OF WOODRIDGE
v.
NAME: *LARRY SALG*
ADDRESS: *3859 Schiller Dr.*
CITY: *Woodridge, Illinois 60517*

The undersigned says that on or about *May 24, 2006,* at or about *9:00 a.m.* the Defendant did unlawfully commit the offense of **Overcrowding** in violation of *2003* **IPMC-404.5** as amended and adopted by reference in Section *8-1J-1(A)* of the *Village of Woodridge* Code, in that said Defendant, the owner* of *34 West Reber St., Woodridge, Illinois, Apartment 4C,* allowed more occupants than permitted by the minimum area requirements of Table 404.5 of the *2003* IPMC as amended and adopted by reference in Section *8-1J-1(A)* of the *Village of Woodridge* Code, in that the requirements allowed *4* persons but there were *11* occupants in the *apartment.*

Karyn Byrne
Complainant

Sworn to and Subscribed before Me
This *17th* Day of *June, 2006*

_____

Notary Public

*Or, occupant.

COMMENT: The 2003 IPMC relies on Table 404.5 to calculate overcrowding based on the size of rooms and the number of occupants.

SAMPLE COMPLAINT

- -

OVERCROWDING—2006 IPMC 404.5

STATE OF *ILLINOIS*
COUNTY OF DUPAGE
VILLAGE OF WOODRIDGE
v.
NAME: *LARRY SALG*
ADDRESS: *3859 Schiller Dr.*
CITY: *Woodridge, Illinois 60517*

The undersigned says that on or about *May 24, 2006*, at or about *9:00 a.m.* the Defendant did unlawfully commit the offense of **Overcrowding** in violation of *2006* **IPMC-404.5** as amended and adopted by reference in Section *8-1J-1(A)* of the *Village of Woodridge* Code, in that said Defendant, the owner* of *34 West Reber St., Woodridge, Illinois, Apartment 4C,* in the opinion of the building official, *Karyn Byrne,* created conditions that endangered the life, health, safety, or welfare of the occupants because of the number of occupants occupying the dwelling unit, being *12,* in that *there was open garbage on the floor of the kitchen and living room, the number of persons made it impossible to safely evacuate during a fire, and the bathroom toilet was overflowing due to the stress on the plumbing by the number of occupants.*

Karyn Byrne
Complainant

Sworn to and Subscribed before Me
This *17th* Day of *June, 2006*

_____
Notary Public

*Or, occupant.

COMMENT: The 2006 IPMC does not rely on square footage to calculate overcrowding based on the size of rooms and the number of occupants. Therefore, the code official must be very specific as to the type of conditions that endanger the life, health, safety, or welfare of the occupants. Because this is a subjective standard, it should be anticipated by the code official that his or her opinion will be challenged in court. Because of the subjective nature, a constitutional challenge to the ordinance on the basis that the ordinance is vague is very likely.

OVERCROWDING (EFFICIENCY)—IPMC 404.6

STATE OF *ILLINOIS*
COUNTY OF DUPAGE
VILLAGE OF WOODRIDGE
v.
NAME: *LARRY SALG*
ADDRESS: *3859 Schiller Dr.*
CITY: *Woodridge, Illinois 60517*

The undersigned says that on or about *May 24, 2006*, at or about *9:00 a.m.* the Defendant did unlawfully commit the offense of **Overcrowding** in violation of *2006* **IPMC-404.6** as amended and adopted by reference in Section *8-1J-1(A)* of the *Village of Woodridge* Code, in that said Defendant, the owner* of *34 West Reber St., Woodridge, Illinois, Apartment 4C,* allowed more occupants than permitted by the minimum area requirements of 2006 IPMC 404.6 as amended and adopted by reference in Section *8-1J-1(A)* of the *Village of Woodridge* Code in that the requirements for an efficiency allowed *3* persons but there were *5* occupants in the efficiency unit.

Karyn Byrne

Complainant

Sworn to and Subscribed before Me
This *17th* Day of *June, 2006*

_____

Notary Public

*Or, occupant.

- -

IMPROPER FOOD PREPARATION AREA—IPMC 404.7

STATE OF *ILLINOIS*
COUNTY OF DUPAGE
VILLAGE OF WOODRIDGE
v.
NAME: *JOHN FINCHAM*
ADDRESS: *34 West Reber St. Apt. 4C*
CITY: *Woodridge, Illinois 60517*

The undersigned says that on or about *May 24, 2006,* at or about *9:00 a.m.* the Defendant did unlawfully commit the offense of **Improper Food Preparation Area** in violation of ***2006 IPMC-404.7*** as amended and adopted by reference in Section *8-1J-1(A)* of the *Village of Woodridge* Code, in that said Defendant, the occupant* of *34 West Reber St., Woodridge, Illinois, Apartment 4C,* did prepare** food *in a bedroom,* a space not suitable for the storage, preparation, and service of foods in a sanitary manner.

Karyn Byrne

Complainant

Sworn to and Subscribed before Me
This *17th* Day of *June, 2006*

_____

Notary Public

  *Or, owner.

**Or, store, prepare, or fail to provide adequate facilities and services for the sanitary disposal of food wastes and refuse.

Plumbing Facilities and Fixture Requirement Violations

CHAPTER 5 covers violations in existing structures which fail to meet the minimum standards for plumbing facilities and fixtures. The chapter prescribes minimum standards for plumbing in dwelling units, rooming houses, hotels, and employees' facilities. It also regulates plumbing in toilet rooms and bathrooms, plumbing systems and fixtures, water systems, sanitary drainage systems, and storm drainage.

IPMC 501　General Violations

FAILURE TO PROVIDE* PLUMBING FACILITIES**—IPMC 501.2

STATE OF *ILLINOIS*
COUNTY OF DUPAGE
VILLAGE OF WOODRIDGE
v.
NAME: *ROBERT MEYER*
ADDRESS: *4617 Carousel St.*
CITY: *Woodridge, Illinois 60517*

The undersigned says that on or about *April 16, 2006*, at or about *3:00 p.m.* the Defendant did unlawfully commit the offense of **Failure to Provide* Plumbing Facilities**** in violation of **2006 IPMC-501.2** as amended and adopted by reference in Section *8-1J-1(A)* of the *Village of Woodridge* Code, in that said Defendant, the owner of *4617 Carousel St., Woodridge, Illinois*, failed to provide* plumbing facilities in violation of 2006 IPMC 502.1 at *4617 Carousel St., Woodridge, Illinois,* in that *there is no bathtub or shower in the dwelling unit.*

Joan Rogers
Complainant

Sworn to and Subscribed before Me
This *30th* Day of *April, 2006*

_____
Notary Public

 *Or, maintain.
**Or, fixtures.

OCCUPYING PREMISES IN VIOLATION OF PLUMBING FACILITIES* REQUIREMENTS—IPMC 501.2

STATE OF *ILLINOIS*
COUNTY OF DUPAGE
VILLAGE OF WOODRIDGE
v.
NAME: *ROBERT MEYER*
ADDRESS: *4617 Carousel St.*
CITY: *Woodridge, Illinois 60517*

The undersigned says that on or about *April 16, 2006*, at or about *3:00 p.m.* the Defendant, *Robert Meyer*, did unlawfully commit the offense **Occupying Premises in Violation of Plumbing Facilities* Requirements** in violation of *2006* **IPMC-501.2** as amended and adopted by reference in Section *8-1J-1(A)* of the *Village of Woodridge* Code; in that said Defendant, the owner-occupant, occupied the premises of *4617 Carousel St., Woodridge, Illinois,* at a time when the premises were in violation of 2006 *IPMC-502.1* in that *there is no bathtub or shower in the dwelling unit.*

Joan Rogers
Complainant

Sworn to and Subscribed before Me
This *30th* Day of *April, 2006*

_____
Notary Public

*Or, fixtures.

SAMPLE COMPLAINT

ALLOWING PERSONS TO OCCUPY PREMISES IN VIOLATION OF PLUMBING FACILITIES*
REQUIREMENTS—IPMC 501.2

STATE OF *ILLINOIS*
COUNTY OF DUPAGE
VILLAGE OF WOODRIDGE
v.
NAME: *ROBERT MEYER*
ADDRESS: *4617 Carousel St.*
CITY: *Woodridge, Illinois 60517*

The undersigned says that on or about *April 16, 2006*, at or about *3:00 p.m.* the Defendant, *Robert Meyer*, did unlawfully commit the offense **Allowing Persons to Occupy Premises in Violation of Plumbing Facilities* Requirements** in violation of *2006* **IPMC-501.2** as amended and adopted by reference in Section *8-1J-1(A)* of the *Village of Woodridge* Code; in that said Defendant, the owner of the premises of *4617 Carousel St., Woodridge, Illinois*, allowed *Keith Steiskal* to occupy the premises at a time when the premises were in violation of 2006 *IPMC-502.1* in that *there is no bathtub or shower in the dwelling unit.*

Joan Rogers
Complainant

Sworn to and Subscribed before Me
This *30th* Day of *April, 2006*

_____
Notary Public

*Or, fixture.

IPMC 502   Required Facilities Violations

TABLE 5-1

REQUIRED PLUMBING FACILITIES

Type of Premises	Required Plumbing Facilities	Section Number
Dwelling Unit	Bathtub or shower	**IPMC 502.1**
	Lavatory (must be in same room as water closet or located in close proximity to door leading directly into the room where water closet is located.)	
	Water closet	
	Kitchen sink (can't be used as substitute for required lavatory)	
Rooming house	One water closet, lavatory, and bathtub or shower for each four rooming units	**IPMC 502.2**
Hotels	Where private water closets, lavatories, and baths not provided: must have one water closet, lavatory, and bathtub or shower access from a public hallway for each 10 occupants	**IPMC 502.3**
Employees' facilities	At least one water closet, lavatory, and drinking facility.	**IPMC 502.4**
	Facilities can be combined with public facilities.	**IPMC 503.3**
	(Drinking facility can be drinking fountain, water cooler, bottled water cooler, or disposable cups next to sink or water dispenser but never in a toilet room or bathroom)	**IPMC 502.4.1**
Employee toilet	Must be not more than one story above or below the working area and the path of travel cannot exceed a distance of 500 feet.	**IPMC 503.3**
Employee toilet where employees work in storage structure or kiosk	Must be located in adjacent structure under same ownership, lease, or control, and travel distance cannot exceed 500 feet from regular working area to facility	**IPMC 503.3 Exception**

FAILURE TO PROVIDE PLUMBING FACILITIES* IN A DWELLING UNIT—IPMC 502.1

STATE OF *ILLINOIS*
COUNTY OF *DUPAGE*
VILLAGE OF WOODRIDGE
v.
NAME: *ROBERT MEYER*
ADDRESS: *4617 Carousel St.*
CITY: *Woodridge, Illinois 60517*

The undersigned says that on or about *April 16, 2006*, at or about *3:00 p.m.* the Defendant did unlawfully commit the offense of **Failure to Provide Plumbing Facilities*** in violation of *2006* **IPMC-501.2** as amended and adopted by reference in Section *8-1J-1(A)* of the *Village of Woodridge* Code, in that said Defendant, the owner of *4617 Carousel St., Woodridge, Illinois,* failed to provide plumbing facilities in violation of 2006 IPMC 502.1 at *4617 Carousel St., Woodridge, Illinois* in that *there is no bathtub or shower in the dwelling unit.*

Joan Rogers
Complainant

Sworn to and Subscribed before Me
This *30th* Day of *April, 2006*

_____

Notary Public

*Or, fixtures.

SAMPLE COMPLAINT

FAILURE TO PROVIDE PLUMBING FACILITIES* IN A DWELLING UNIT—IPMC 502.1

STATE OF *ILLINOIS*
COUNTY OF DUPAGE
VILLAGE OF WOODRIDGE
v.
NAME: *ROBERT MEYER*
ADDRESS: *4617 Carousel St.*
CITY: *Woodridge, Illinois 60517*

The undersigned says that on or about *April 16, 2006*, at or about *3:00 p.m.* the Defendant did unlawfully commit the offense of **Failure to Provide Plumbing Facilities*** in violation of ***2006* IPMC-502.1** as amended and adopted by reference in Section *8-1J-1(A)* of the *Village of Woodridge* Code, in that said Defendant, the owner of *4617 Carousel St., Woodridge, Illinois*, failed to provide plumbing facilities* in violation of 2006 IPMC-502.1 in that *the kitchen sink is being used as a required lavatory.*

Joan Rogers

Complainant

Sworn to and Subscribed before Me
This *30th* Day of *April, 2006*

_____

Notary Public

*Or, fixtures.

FAILURE TO MAINTAIN PLUMBING FIXTURES IN A DWELLING UNIT—IPMC 502.1

STATE OF *ILLINOIS*
COUNTY OF DUPAGE
VILLAGE OF WOODRIDGE
v.
NAME: *ROBERT MEYER*
ADDRESS: *4617 Carousel St.*
CITY: *Woodridge, Illinois 60517*

The undersigned says that on or about *April 16, 2006,* at or about *3:00 p.m.* the Defendant did unlawfully commit the offense of **Failure to Maintain Plumbing Fixtures** in violation of **2006 IPMC-501.2** as amended and adopted by reference in Section *8-1J-1(A)* of the *Village of Woodridge* Code, in that said Defendant, the owner of *4617 Carousel St., Woodridge, Illinois,* failed to provide plumbing fixtures in a sanitary, safe working condition in that *the water closet is clogged and water is overflowing onto the floor.*

Joan Rogers
Complainant

Sworn to and Subscribed before Me
This *30th* Day of *April, 2006*

_____

Notary Public

SAMPLE COMPLAINT

FAILURE TO PROVIDE PLUMBING FACILITIES IN A ROOMING HOUSE—IPMC 502.2

STATE OF *ILLINOIS*
COUNTY OF DUPAGE
VILLAGE OF WOODRIDGE
v.
NAME: *KEITH STEISKAL*
ADDRESS: *4617 Carousel St.*
CITY: *Woodridge, Illinois 60517*

The undersigned says that on or about *May 24, 2006*, at or about *3:00 p.m.* the Defendant did unlawfully commit the offense of **Failure to Provide Plumbing Facilities in a Rooming House** in violation of *2006* **IPMC-502.2** as amended and adopted by reference in Section *8-1J-1(A)* of the *Village of Woodridge* Code, in that said Defendant, the owner of *2206 W. Candlewood Ct., Woodridge, Illinois*, a rooming house, failed to provide plumbing facilities in said premises in that *there is only one water closet for six rooming units.*

Joan Rogers
Complainant

Sworn to and Subscribed before Me
This *30th* Day of *April, 2006*

_____
Notary Public

COMMENT: There must be one water closet, lavatory, and bathtub or shower for each four rooming units.

FAILURE TO PROVIDE PLUMBING FACILITIES IN A HOTEL—IPMC 502.3

STATE OF *ILLINOIS*
COUNTY OF DUPAGE
VILLAGE OF WOODRIDGE
v.
NAME: *KEITH STEISKAL*
ADDRESS: *4617 Carousel St.*
CITY: *Woodridge, Illinois 60517*

The undersigned says that on or about *May 24, 2006*, at or about *3:00 p.m.* the Defendant did unlawfully commit the offense of **Failure to Provide Plumbing Facilities in a Hotel** in violation of *2006* **IPMC-502.3** as amended and adopted by reference in Section *8-1J-1(A)* of the *Village of Woodridge* Code, in that said Defendant, the owner of *2210 W. Candlewood Ct., Woodridge, Illinois*, a hotel without private water closets, lavatories, and baths, failed to provide plumbing facilities in said premises in that *there is only one shower for every twenty occupants.*

Joan Rogers
Complainant

Sworn to and Subscribed before Me
This *30th* Day of *April, 2006*

_____
Notary Public

COMMENT: There must be one water closet, lavatory, and bathtub or shower for every ten occupants.

SAMPLE COMPLAINT

FAILURE TO PROVIDE DRINKING FACILITIES*—IPMC 502.4

STATE OF *ILLINOIS*
COUNTY OF DUPAGE
VILLAGE OF WOODRIDGE
v.
NAME: *KYLE'S CHICKEN SHACK, INC.*
ADDRESS: *109 Fremont St.*
CITY: *Woodridge, Illinois 60517*

The undersigned says that on or about *April 2, 2006*, at or about *3:00 p.m.* the Defendant did unlawfully commit the offense of **Failure to Provide Drinking Facilities*** in violation of **2006 IPMC-502.4** as amended and adopted by reference in Section *8-1J-1(A)* of the *Village of Woodridge* Code, in that said Defendant, the owner of *109 Fremont St., Woodridge, Illinois,* failed to provide drinking facilities* for the employees at the *restaurant* located on the premises in that *there is no drinking fountain, water cooler, bottled water cooler, or disposable cups next to a sink or water dispenser.*

Joan Rogers
Complainant

Sworn to and Subscribed before Me
This *30th* Day of *April, 2006*

_____

Notary Public

*Or, water closet or lavatory.

IPMC 503   Toilet Room Violations

IPMC 503.1—Privacy

Sample Complaint—Failure to Provide Privacy in a Toilet Room   203

IPMC 503.2—Location

Sample Complaint—Improper Location of a Toilet Room in a Hotel Unit   204

IPMC 503.3—Location of Employee Toilet Facilities

Sample Complaint—Improper Location of Employees' Toilet Facility   205

IPMC 503.4—Floor Surface

Sample Complaint—Improper Floor Surface in a Toilet Room   206

FAILURE TO PROVIDE PRIVACY IN A TOILET ROOM*—IPMC 503.1

STATE OF *ILLINOIS*
COUNTY OF DUPAGE
VILLAGE OF WOODRIDGE
v.
NAME: *KEITH STEISKAL*
ADDRESS: *4617 Carousel St.*
CITY: *Woodridge, Illinois 60517*

The undersigned says that on or about *May 24, 2006*, at or about *3:00 p.m.* the Defendant did unlawfully commit the offense of **Failure to Provide Privacy in a Toilet Room*** in violation of ***2006* IPMC-503.1** as amended and adopted by reference in Section *8-1J-1(A)* of the *Village of Woodridge* Code in that said Defendant, the owner of *2206 W. Candlewood Ct., Woodridge, Illinois*, a rooming house, failed to provide privacy in a toilet room* in said premises in that *there is no interior locking device for the shared toilet room.***

Joan Rogers
Complainant

Sworn to and Subscribed before Me
This *30th* Day of *April, 2006*

_____

Notary Public

  *Or, bathroom.
**Or, the toilet room/bathroom is the only passageway to a hall or other space.

IMPROPER LOCATION OF A TOILET ROOM* IN A HOTEL UNIT**—IPMC 503.2

STATE OF *ILLINOIS*
COUNTY OF DUPAGE
VILLAGE OF WOODRIDGE
v.
NAME: *KEITH STEISKAL*
ADDRESS: *4617 Carousel St.*
CITY: *Woodridge, Illinois 60517*

The undersigned says that on or about *May 24, 2006*, at or about *3:00 p.m.* the Defendant did unlawfully commit the offense of **Improper Location of a Toilet Room* in a Hotel Unit**** in violation of *2006 IPMC-503.2* as amended and adopted by reference in Section *8-1J-1(A)* of the *Village of Woodridge* Code, in that said Defendant, the owner of *2210 W. Candlewood Ct., Woodridge, Illinois*, a hotel** without private water closets, lavatories and baths, failed to provide a toilet room* which was properly located in that *occupants on the first floor must go up two flights of stairs to reach a toilet room.****

Joan Rogers
Complainant

Sworn to and Subscribed before Me
This *17th* Day of *June, 2006*

_____
Notary Public

  *Or, bathroom.
 **Or, rooming units, dormitory units or housekeeping units.
***Or, in that occupants do not have access from a common hall or passageway.

SAMPLE COMPLAINT

IMPROPER LOCATION OF EMPLOYEES' TOILET FACILITY—IPMC 503.3

STATE OF *ILLINOIS*
COUNTY OF DUPAGE
VILLAGE OF WOODRIDGE
v.
NAME: *KYLE'S CHICKEN SHACK, INC.*
ADDRESS: *109 Fremont St.*
CITY: *Woodridge, Illinois 60517*

The undersigned says that on or about *April 2, 2006*, at or about *3:00 p.m.* the Defendant did unlawfully commit the offense of **Improper Location of Employees' Toilet Facility** in violation of *2006* **IPMC-503.3** as amended and adopted by reference in Section *8-1J-1(A)* of the *Village of Woodridge* Code, in that said Defendant, the owner of *109 Fremont St., Woodridge, Illinois*, failed to provide a properly located toilet facility for the employees in that *the employees have to travel 1,000 feet to reach a toilet facility in a business not owned by the Defendant.*

Joan Rogers
Complainant

Sworn to and Subscribed before Me
This *30th* Day of *April, 2006*

_____
Notary Public

COMMENT: Toilet facilities must be not more than one story above or below the working area, and the path of travel cannot exceed a distance of 500 feet unless the employees work in a storage structure or kiosk, in which case they must be located in an adjacent structure under the same ownership, lease, or control, and the travel distance cannot exceed 500 feet from regular working area to facility.

IMPROPER FLOOR SURFACE IN A TOILET ROOM—IPMC 503.4

STATE OF *ILLINOIS*
COUNTY OF DUPAGE
VILLAGE OF WOODRIDGE
v.
NAME: *KYLE'S CHICKEN SHACK, INC.*
ADDRESS: *109 Fremont St.*
CITY: *Woodridge, Illinois 60517*

The undersigned says that on or about *April 2, 2006*, at or about *3:00 p.m.* the Defendant did unlawfully commit the offense of **Improper Floor Surface in a Toilet Room** in violation of **2006 IPMC-503.4** as amended and adopted by reference in Section *8-1J-1(A)* of the *Village of Woodridge* Code, in that said Defendant, the owner of *109 Fremont St., Woodridge, Illinois*, failed to provide a proper floor surface in a toilet room in the building, which does not contain dwelling units, in that the floor surface was made out of *wood*, not being a nonabsorbent material, making it difficult to keep it in a clean and sanitary condition.

Joan Rogers
Complainant

Sworn to and Subscribed before Me
This *30th* Day of *April, 2006*

_____

Notary Public

IPMC 504   Plumbing System and Fixture Violations

SAMPLE COMPLAINT

- -

FAILURE TO MAINTAIN* PLUMBING FIXTURES—IPMC 504.1

STATE OF *ILLINOIS*
COUNTY OF DUPAGE
VILLAGE OF WOODRIDGE
v.
NAME: *ROBERT MEYER*
ADDRESS: *4617 Carousel St.*
CITY: *Woodridge, Illinois 60517*

The undersigned says that on or about *April 16, 2006*, at or about *3:00 p.m.* the Defendant did unlawfully commit the offense of **Failure to Maintain Plumbing Fixtures** in violation of **2006 IPMC-504.1** as amended and adopted by reference in Section *8-1J-1(A)* of the *Village of Woodridge* Code, in that said Defendant, the owner of *4617 Carousel St., Woodridge, Illinois*, failed to maintain* plumbing fixtures in a sanitary, safe, and functional condition in that *the water closet is clogged and water is overflowing onto the floor.*

Joan Rogers
Complainant

Sworn to and Subscribed before Me
This *30th* Day of *April, 2006*

_____
Notary Public

*Or, properly install or keep free from obstructions, leaks, and defects and be capable of performing the function for which it is designed.

IMPROPER FIXTURE CLEARANCE—IPMC 504.2

STATE OF *ILLINOIS*
COUNTY OF DUPAGE
VILLAGE OF WOODRIDGE
v.
NAME: *ROBERT MEYER*
ADDRESS: *4617 Carousel St.*
CITY: *Woodridge, Illinois 60517*

The undersigned says that on or about *April 16, 2006*, at or about *3:00 p.m.* the Defendant, the owner of *4617 Carousel St., Woodridge, Illinois*, did unlawfully commit the offense of **Improper Fixture Clearance** in violation of *2006* **IPMC-504.2** as amended and adopted by reference in Section *8-1J-1(A)* of the *Village of Woodridge* Code, in that *the toilet in the bathroom is installed so close to the sink that it is difficult to use and clean.*

Joan Rogers
Complainant

Sworn to and Subscribed before Me
This *30th* Day of *April, 2006*

_____

Notary Public

PLUMBING SYSTEM HAZARD—IPMC 504.3

STATE OF *ILLINOIS*

COUNTY OF DUPAGE

VILLAGE OF WOODRIDGE

v.

NAME: *ROBERT MEYER*

ADDRESS: *4617 Carousel St.*

CITY: *Woodridge, Illinois 60517*

The undersigned says that on or about *April 16, 2006,* at or about *3:00 p.m.* the Defendant did unlawfully commit the offense of **Plumbing System Hazard** in violation of *2006* **IPMC-504.3** as amended and adopted by reference in Section *8-1J-1(A)* of the *Village of Woodridge* Code, in that said Defendant, the owner of *4617 Carousel St., Woodridge, Illinois,* did after being ordered by the code official, *Joan Rogers,* to correct a plumbing system hazard to the structure and the occupants on the premises, *being a failure to vent* the upstairs toilet,* did fail to make such correction.

Joan Rogers

Complainant

Sworn to and Subscribed before Me

This *30th* Day of *April, 2006*

_____

Notary Public

*Or, failure to provide adequate service, improper cross connection, backsiphonage, improper installation, deterioration, or damage.

IPMC 505   Water System Violations

IMPROPER WATER SYSTEM CONNECTION—IPMC 505.1

STATE OF *ILLINOIS*
COUNTY OF DUPAGE
VILLAGE OF WOODRIDGE
v.
NAME: *ROBERT MEYER*
ADDRESS: *4617 Carousel St.*
CITY: *Woodridge, Illinois 60517*

The undersigned says that on or about *April 16, 2006*, at or about *3:00 p.m.* the Defendant did unlawfully commit the offense of **Improper Water System Connection** in violation of **2006 IPMC-505.1** as amended and adopted by reference in Section *8-1J-1(A)* of the *Village of Woodridge* Code, in that said Defendant, the owner of *4617 Carousel St., Woodridge, Illinois*, improperly connected the laundry facilities* to the public** water system in that *the drain for the washing machine is too far away from the soil stack.*

Joan Rogers
Complainant

Sworn to and Subscribed before Me
This *30th* Day of *April, 2006*

_____
Notary Public

 *Or, sink, lavatory, bathtub, shower, drinking fountain, water closet, or other plumbing facility.
**Or, private.

SAMPLE COMPLAINT

CONTAMINATION OF THE WATER SUPPLY—IPMC 505.2

STATE OF *ILLINOIS*
COUNTY OF DUPAGE
VILLAGE OF WOODRIDGE
v.
NAME: *ROBERT MEYER*
ADDRESS: *4617 Carousel St.*
CITY: *Woodridge, Illinois 60517*

The undersigned says that on or about *April 16, 2006*, at or about *3:00 p.m.* the Defendant did unlawfully commit the offense of **Contamination of the Water Supply** in violation of **2006 IPMC-505.2** as amended and adopted by reference in Section *8-1J-1(A)* of the *Village of Woodridge* Code, in that said Defendant, the owner of *4617 Carousel St., Woodridge, Illinois*, failed to maintain the water supply free from contamination in that *the faucet for the bathroom sink is located below the flood-level rim of the fixture, thereby establishing an improper cross-connection.*

Joan Rogers

Complainant

Sworn to and Subscribed before Me
This *30th* Day of *April, 2006*

_____

Notary Public

IMPROPER VACUUM BREAKER—IPMC 505.2

STATE OF *ILLINOIS*
COUNTY OF DUPAGE
VILLAGE OF WOODRIDGE
v.
NAME: *KYLE'S CHICKEN SHACK, INC.*
ADDRESS: *109 Fremont St.*
CITY: *Woodridge, Illinois 60517*

The undersigned says that on or about *April 16, 2006*, at or about *3:00 p.m.* the Defendant did unlawfully commit the offense of **Improper Vacuum Breaker** in violation of ***2006 IPMC-505.2*** as amended and adopted by reference in Section *8-1J-1(A)* of the *Village of Woodridge* Code, in that said Defendant, the owner of *109 Fremont St., Woodridge, Illinois*, had a *janitor sink faucet*** without an approved atmospheric-type vacuum breaker.****

Joan Rogers
Complainant

Sworn to and Subscribed before Me
This *30th* Day of *April, 2006*

_____

Notary Public

 *Or, shampoo basin faucet, hose bib, or faucet to which a hose is attached and left in place.
**Or, approved permanently attached hose connection vacuum breaker.

SAMPLE COMPLAINT

IMPROPER WATER SUPPLY SYSTEM—IPMC 505.3

STATE OF *ILLINOIS*
COUNTY OF DUPAGE
VILLAGE OF WOODRIDGE
v.
NAME: *ROBERT MEYER*
ADDRESS: *4617 Carousel St.*
CITY: *Woodridge, Illinois 60517*

The undersigned says that on or about *April 16, 2006*, at or about *3:00 p.m.* the Defendant did unlawfully commit the offense of **Improper Water Supply System** in violation of **2006 IPMC-505.3** as amended and adopted by reference in Section *8-1J-1(A)* of the *Village of Woodridge* Code, in that said Defendant, the owner of *4617 Carousel St., Woodridge, Illinois,* failed to maintain* the water supply system for the premises so as to provide a supply of water in sufficient volume and at pressures adequate to enable the fixtures to function property, safely, and free from defects and leaks in *that the water pipes leading into the kitchen are leaking.*

Joan Rogers
Complainant

Sworn to and Subscribed before Me
This *30th* Day of *April, 2006*

_____
Notary Public

*Or, install.

IMPROPER WATER HEATING FACILITY—IPMC 505.4

STATE OF *ILLINOIS*
COUNTY OF DUPAGE
VILLAGE OF WOODRIDGE
v.
NAME: *ROBERT MEYER*
ADDRESS: *4617 Carousel St.*
CITY: *Woodridge, Illinois 60517*

The undersigned says that on or about *April 16, 2006,* at or about *3:00 p.m.* the Defendant did unlawfully commit the offense of **Improper Water Heating Facility** in violation of **2006 IPMC-505.4** as amended and adopted by reference in Section *8-1J-1(A)* of the *Village of Woodridge* Code, in that said Defendant, the owner of *4617 Carousel St., Woodridge, Illinois,* failed to maintain* the water heating facility so as to provide an adequate amount of water to the sink, lavatory, bathtub, shower and laundry facility at a temperature of 110°F in that *water is boiling from over the top of the hot water heater.*

Joan Rogers
Complainant

Sworn to and Subscribed before Me
This *30th* Day of *April, 2006*

_____
Notary Public

*Or, install.

SAMPLE COMPLAINT

IMPROPER WATER HEATER LOCATION—IPMC 505.4

STATE OF *ILLINOIS*
COUNTY OF DUPAGE
VILLAGE OF WOODRIDGE
v.
NAME: *ROBERT MEYER*
ADDRESS: *4617 Carousel St.*
CITY: *Woodridge, Illinois 60517*

The undersigned says that on or about *April 16, 2006*, at or about *3:00 p.m.* the Defendant did unlawfully commit the offense of **Improper Water Heater Location** in violation of ***2006 IPMC-505.4*** as amended and adopted by reference in Section *8-1J-1(A)* of the *Village of Woodridge* Code, in that said Defendant, the owner of *4617 Carousel St., Woodridge, Illinois,* had a gas-burning water heater in a *bathroom** without adequate combustion air.

Joan Rogers
Complainant

Sworn to and Subscribed before Me
This *30th* Day of *April, 2006*

_____

Notary Public

*Or, toilet room, bedroom, or other occupied room normally kept closed.

IMPROPER WATER HEATER VALVE AND PIPE—IPMC 505.4

STATE OF *ILLINOIS*
COUNTY OF DUPAGE
VILLAGE OF WOODRIDGE
v.
NAME: *ROBERT MEYER*
ADDRESS: *4617 Carousel St.*
CITY: *Woodridge, Illinois 60517*

The undersigned says that on or about *April 16, 2006*, at or about *3:00 p.m.* the Defendant did unlawfully commit the offense of **Improper Water Heater Valve and Pipe** in violation of *2006* **IPMC-505.4** as amended and adopted by reference in Section *8-1J-1(A)* of the *Village of Woodridge* Code, in that said Defendant, the owner of *4617 Carousel St., Woodridge, Illinois,* installed* a water heater without an approved pressure-relief valve and relief valve discharge pipe.

Joan Rogers
Complainant

Sworn to and Subscribed before Me
This *30th* Day of *April, 2006*

_____
Notary Public

*Or, maintained.

IPMC 506   Sanitary Drainage System Violations

FAILURE TO CONNECT TO SANITARY DRAINAGE SYSTEM—IPMC 506.1

STATE OF *ILLINOIS*
COUNTY OF DUPAGE
VILLAGE OF WOODRIDGE
v.
NAME: *ROBERT MEYER*
ADDRESS: *4617 Carousel St.*
CITY: *Woodridge, Illinois 60517*

The undersigned says that on or about *April 16, 2006*, at or about *3:00 p.m.* the Defendant did unlawfully commit the offense of **Failure to Connect to Sanitary Drainage System** in violation of *2006* **IPMC-506.1** as amended and adopted by reference in Section *8-1J-1(A)* of the *Village of Woodridge* Code, in that said Defendant, the owner of *4617 Carousel St., Woodridge, Illinois,* did fail to connect a water closet to the public sewer system* in that *the drainage pipe leads directly out the side of the house onto the side yard.*

Joan Rogers
Complainant

Sworn to and Subscribed before Me
This *30th* Day of *April, 2006*

_____

Notary Public

*Or, approved private sewage disposal system.

SAMPLE COMPLAINT

FAILURE TO MAINTAIN SANITARY DRAINAGE SYSTEM—IPMC 506.2

STATE OF *ILLINOIS*
COUNTY OF DUPAGE
VILLAGE OF WOODRIDGE
v.
NAME: *ROBERT MEYER*
ADDRESS: *4617 Carousel St.*
CITY: *Woodridge, Illinois 60517*

The undersigned says that on or about *April 16, 2006*, at or about *3:00 p.m.* the Defendant did unlawfully commit the offense of **Failure to Maintain Sanitary Drainage System** in violation of *2006* **IPMC-506.2** as amended and adopted by reference in Section *8-1J-1(A)* of the *Village of Woodridge* Code, in that said Defendant, the owner of *4617 Carousel St., Woodridge, Illinois*, did fail to maintain a waste line* so that it functioned properly and was kept free from obstructions, leaks, and defects in that *there is a backup of sewage from a drain in the basement.*

Joan Rogers
Complainant

Sworn to and Subscribed before Me
This *30th* Day of *April, 2006*

_____
Notary Public

*Or, plumbing stack, vent, or sewer line.

IPMC 507   Storm Drainage Violations

IPMC 507.1—General

SAMPLE COMPLAINT
- -

IMPROPER DRAINAGE—IPMC 507.1

STATE OF *ILLINOIS*
COUNTY OF DUPAGE
VILLAGE OF WOODRIDGE
v.
NAME: *ROBERT MEYER*
ADDRESS: *4617 Carousel St.*
CITY: *Woodridge, Illinois 60517*

The undersigned says that on or about *November 18, 2006*, at or about *3:00 p.m.* the Defendant did unlawfully commit the offense of **Improper Drainage** in violation of *2006* **IPMC-507.1** as amended and adopted by reference in Section *8-1J-1(A)* of the *Village of Woodridge* Code, in that said Defendant, the owner of *4617 Carousel St., Woodridge, Illinois*, did allow storm water to drain from *his driveway* onto the public street and sidewalk* so as to create a public nuisance in that *discharged sump pump water freezes, creating a hazard for pedestrians and vehicles.*

Joan Rogers
Complainant

Sworn to and Subscribed before Me
This *7th* Day of *December, 2006*

_____
Notary Public

*Or, roofs, paved areas, courts and other open areas.

Mechanical and Electrical Violations

CHAPTER 6 covers violations in existing structures which fail to meet the minimum standards for mechanical and electrical requirements. The chapter prescribes minimum standards for heating in dwellings and work spaces. It also regulates mechanical equipment, electrical facilities and equipment, elevators, escalators, dumbwaiters, and duct systems.

IPMC 601   General Violations

IPMC 601.2—Responsibility

Sample Complaint—Failure to Provide
Mechanical Equipment   227

Sample Complaint—Occupying Premises in
Violation of Mechanical Equipment
Requirements   228

Sample Complaint—Allowing Persons to
Occupy Premises in Violation of Heating
Facilities Requirements   229

FAILURE TO PROVIDE* MECHANICAL** EQUIPMENT***—IPMC 601.2

STATE OF *ILLINOIS*
COUNTY OF DUPAGE
VILLAGE OF WOODRIDGE
v.
NAME: *PAT FOLEY*
ADDRESS: *7760 County Line Lane*
CITY: *Woodridge, Illinois 60517*

The undersigned says that on or about *December 26, 2006*, at or about *10:00 a.m.* the Defendant did unlawfully commit the offense of **Failure to Provide* Mechanical** Equipment***** in violation of ***2006 IPMC-601.2*** as amended and adopted by reference in Section *8-1J-1(A)* of the *Village of Woodridge* Code, in that said Defendant, the owner of *7660 Chicago Avenue, Woodridge, Illinois*, failed to provide* mechanical** equipment*** on the premises in violation of *2006 IPMC 603.1* as amended and adopted by reference in Section *8-1J-1(A)* of the *Village of Woodridge* Code, in that *there is no functioning hot water heater in the dwelling.*

Karyn Byrne
Complainant

Sworn to and Subscribed before Me
This *27th* Day of *December, 2006*

_____
Notary Public

   *Or, maintain.
  **Or, heating or electrical.
***Or, facilities.

OCCUPYING PREMISES IN VIOLATION OF MECHANICAL* EQUIPMENT** REQUIREMENTS—IPMC 601.2

STATE OF *ILLINOIS*
COUNTY OF DUPAGE
VILLAGE OF WOODRIDGE
v.
NAME: *PAT FOLEY*
ADDRESS: *7760 County Line Lane*
CITY: *Woodridge, Illinois 60517*

The undersigned says that on or about *December 26, 2006*, at or about *10:00 a.m.* the Defendant did unlawfully commit the offense of **Occupying Premises in Violation of Mechanical* Equipment** Requirements** in violation of *2006* **IPMC-601.2** as amended and adopted by reference in Section *8-1J-1(A)* of the *Village of Woodridge* Code, in that said Defendant, the owner-occupant of *7660 Chicago Avenue., Woodridge, Illinois*, occupied said premises at a time when the premises were in violation of *2006* IPMC *603.1* as amended and adopted by reference in Section *8-1J-1(A)* of the *Village of Woodridge* Code, in that *there was no functioning hot water heater in the dwelling.*

Karyn Byrne

Complainant

Sworn to and Subscribed before Me
This *27th* Day of *December, 2006*

_____

Notary Public

 *Or, heating or electrical.
**Or, facilities.

SAMPLE COMPLAINT

ALLOWING PERSONS TO OCCUPY PREMISES IN VIOLATION OF HEATING* FACILITIES** REQUIREMENTS—IPMC 601.2

STATE OF *ILLINOIS*
COUNTY OF DUPAGE
VILLAGE OF WOODRIDGE
v.
NAME: *PAT FOLEY*
ADDRESS: *7760 County Line Lane*
CITY: *Woodridge, Illinois 60517*

The undersigned says that on or about *December 26, 2006,* at or about *10:00 a.m.* the Defendant did unlawfully commit the offense of **Allowing Persons to Occupy Premises in Violation of Heating* Facility** Requirements** in violation of *2006* **IPMC-601.2** as amended and adopted by reference in Section *8-1J-1(A)* of the *Village of Woodridge* Code, in that said Defendant, the owner of *7660 Chicago Avenue, Woodridge, Illinois,* allowed persons to occupy said premises at a time when the premises were in violation of *2006* IPMC *602.3* as amended and adopted by reference in Section *8-1J-1(A)* of the *Village of Woodridge* Code in that *the defendant failed to furnish heat to the occupants of the dwelling in that the temperature in the dwelling was 52°F.*

Karyn Byrne

Complainant

Sworn to and Subscribed before Me
This *27th* Day of *December, 2006*

_____

Notary Public

 *Or, mechanical or electrical.
**Or, equipment.

IPMC 602   Heating Facilities

TABLE 6-1

HEATING REQUIREMENTS

Location	Room Temperature	Code Section
Habitable rooms, bathrooms, and toilet rooms in dwellings	Must have heating facilities capable of maintaining room temperature of 68°F (20°C) **Exception:** Where average monthly temperature is above 30°F (−1°C), minimum temperature to be maintained is 65°F (18°C)	602.2
Dwelling unit, rooming unit, dormitory or guestroom on terms	Not less than 68°F (20°C) from (date*) to (date*) or, where average monthly temperature is above 30°F (−1°C), then not less than 65°F (18°C) **Exception:** When outdoor temperature is below the winter outdoor design temperature for the locality minimum room temperature not required if heating system is operating at full design capacity	602.3
Indoor occupiable work spaces	Not less than 65°F (18°C) from (date*) to (date*) **Exception:** processing, storage, and operation areas that require cooling or special temperatures or areas in which persons primarily engage in vigorous physical activity	602.4

*Insert dates for when heat is required, e.g., November 1 to April 30.

FAILURE TO PROVIDE HEATING FACILITIES—IPMC 602.1

STATE OF *ILLINOIS*
COUNTY OF DUPAGE
VILLAGE OF WOODRIDGE
v.
NAME: *PAT FOLEY*
ADDRESS: *7760 County Line Lane*
CITY: *Woodridge, Illinois 60517*

The undersigned says that on or about *December 26, 2006*, at or about *10:00 a.m.* the Defendant did unlawfully commit the offense of **Failure to Provide Heating Facilities** in violation of *2006* **IPMC-602.1** as amended and adopted by reference in Section *8-1J-1(A)* of the *Village of Woodridge* Code, in that said Defendant, the owner of *7660 Chicago Avenue, Woodridge, Illinois*, failed to provide heating facilities at said premises in that *there was no functioning furnace in the dwelling.*

Karyn Byrne
Complainant

Sworn to and Subscribed before Me
This *27th* Day of *December, 2006*

_____
Notary Public

SAMPLE COMPLAINT

- -

FAILURE TO PROVIDE PROPER HEATING FACILITIES IN A RESIDENCE—IPMC 602.2

STATE OF *ILLINOIS*
COUNTY OF DUPAGE
VILLAGE OF WOODRIDGE
v.
NAME: *PAT FOLEY*
ADDRESS: *7760 County Line Lane*
CITY: *Woodridge, Illinois 60517*

The undersigned says that on or about *December 26, 2006,* at or about *10:00 a.m.* the Defendant did unlawfully commit the offense of **Failure to Provide Proper Heating Facilities in a Residence** in violation of *2006* **IPMC-602.2** as amended and adopted by reference in Section *8-1J-1(A)* of the *Village of Woodridge* Code, in that said Defendant, the owner of *7660 Chicago Avenue, Woodridge, Illinois,* failed to provide proper heating facilities at said premises in that *the furnace in the dwelling* could only maintain a room temperature of *58°F in the bathroom* and not the required 68°F.*

Karyn Byrne

Complainant

Sworn to and Subscribed before Me
This *27th* Day of *December, 2006*

_____

Notary Public

*In areas where the average monthly temperature is above 30°F (−1°C), the minimum
 temperature would be 65°F (18°C).

COMMENT: Local authorities may amend this section sometimes to provide for higher minimum temperatures. The code official should check the local jurisdiction's requirements.

- -

FAILURE TO PROVIDE PROPER HEAT SUPPLY IN A RESIDENCE—IPMC 602.3

STATE OF *ILLINOIS*
COUNTY OF DUPAGE
VILLAGE OF WOODRIDGE
v.
NAME: *PAT FOLEY*
ADDRESS: *7760 County Line Lane*
CITY: *Woodridge, Illinois 60517*

The undersigned says that on or about *December 26, 2006*, at or about *10:00 a.m.* the Defendant did unlawfully commit the offense of **Failure to Provide Proper Heat Supply in a Residence** in violation of ***2006* IPMC-602.3** as amended and adopted by reference in Section *8-1J-1(A)* of the *Village of Woodridge* Code, in that said Defendant, the owner of *7660 Chicago Avenue, Woodridge, Illinois*, who rented* the premises to another, failed to provide a proper heat supply in the dwelling unit** at said premises in that the room temperature was *50°*F in the *bedroom*.

Karyp Byrne

Complainant

Sworn to and Subscribed before Me
This *27th* Day of *December, 2006*

_____

Notary Public

 *Or, leased or lets.
**Or, rooming unit, dormitory, or guestroom on terms.

SAMPLE COMPLAINT

FAILURE TO PROVIDE PROPER HEAT SUPPLY IN AN OCCUPIABLE WORK SPACE—IPMC 602.4

STATE OF *ILLINOIS*
COUNTY OF DUPAGE
VILLAGE OF WOODRIDGE
v.
NAME: *PAT FOLEY*
ADDRESS: *7760 County Line Lane*
CITY: *Woodridge, Illinois 60517*

The undersigned says that on or about *December 26, 2006*, at or about *10:00 a.m.* the Defendant did unlawfully commit the offense of **Failure to Provide Proper Heat Supply in an Occupiable Work Space** in violation of *2006* **IPMC-602.4** as amended and adopted by reference in Section *8-1J-1(A)* of the *Village of Woodridge* Code, in that said Defendant, the owner of *4000 Quincy St., Woodridge, Illinois,* failed to provide a proper heat supply in the occupiable work space located therein, being the *office area* at said premises in that the room temperature was *56°F.*

Karyn Byrne
Complainant

Sworn to and Subscribed before Me
This *3rd* Day of *January, 2007*

_____

Notary Public

ILLUSTRATION 6-1

ROOM TEMPERATURE MEASUREMENT—IPMC 602.5

Room temperatures must be measured 3 feet (914 mm) above the floor near the center of the room and 2 feet (610 mm) inward from the center of each exterior wall. Failure to perform the measurement properly will result in an invalid measurement. Therefore, there will be insufficient evidence to prove the case. **IPMC** 602.5

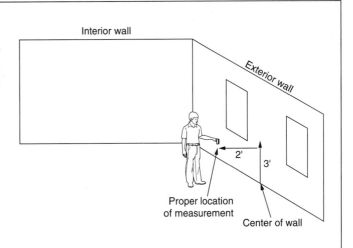

IPMC 603   Mechanical Equipment Violations

SAMPLE COMPLAINT

FAILURE TO MAINTAIN* PROPER MECHANICAL APPLIANCE—IPMC 603.1

STATE OF *ILLINOIS*
COUNTY OF DUPAGE
VILLAGE OF WOODRIDGE
v.
NAME: *PAT FOLEY*
ADDRESS: *7760 County Line Lane*
CITY: *Woodridge, Illinois 60517*

The undersigned says that on or about *December 26, 2006*, at or about *10:00 a.m.* the Defendant did unlawfully commit the offense of **Failure to Maintain* Mechanical Appliance** in violation of *2006* **IPMC-603.1** as amended and adopted by reference in Section *8-1J-1(A)* of the *Village of Woodridge* Code, in that said Defendant, the owner of *7660 Chicago Avenue, Woodridge, Illinois*, failed to maintain* a proper cooking appliance** in a safe working condition capable of performing the intended function in that *the gas stove burners do not ignite automatically, thereby spewing natural gas into the kitchen.*

Karyn Byrne

Complainant

Sworn to and Subscribed before Me
This *3rd* Day of *January, 2007*

_____

Notary Public

 *Or, install.
**Or, mechanical fireplace, solid-burning appliance, or water heating appliance.

SAMPLE COMPLAINT

FAILURE TO REMOVE COMBUSTION PRODUCTS—IPMC 603.2

STATE OF *ILLINOIS*
COUNTY OF DUPAGE
VILLAGE OF WOODRIDGE
v.
NAME: *PAT FOLEY*
ADDRESS: *7760 County Line Lane*
CITY: *Woodridge, Illinois 60517*

The undersigned says that on or about *December 26, 2006,* at or about *10:00 a.m.* the Defendant did unlawfully commit the offense of **Failure to Remove Combustion Products** in violation of *2006* **IPMC-603.2** as amended and adopted by reference in Section *8-1J-1(A)* of the *Village of Woodridge* Code, in that said Defendant, the owner of *7660 Chicago Avenue, Woodridge, Illinois,* failed to connect a *furnace*, a fuel-burning appliance,* to an approved chimney** *in that the vent from the furnace went into the crawl space instead of the chimney.*

Karyn Byrne

Complainant

Sworn to and Subscribed before Me
This *3rd* Day of *January, 2007*

_____
Notary Public

 *Or, equipment.
**Or, vent.

EXCEPTION: Fuel-burning equipment and appliances which are labeled for unvented operation.

SAMPLE COMPLAINT

FAILURE TO PROVIDE REQUIRED CLEARANCE—IPMC 603.3

STATE OF *ILLINOIS*
COUNTY OF DUPAGE
VILLAGE OF WOODRIDGE
v.
NAME: *PAT FOLEY*
ADDRESS: *7760 County Line Lane*
CITY: *Woodridge, Illinois 60517*

The undersigned says that on or about *December 26, 2006*, at or about *10:00 a.m.* the Defendant did unlawfully commit the offense of **Failure to Provide Required Clearance** in violation of **2006 IPMC-603.3** as amended and adopted by reference in Section *8-1J-1(A)* of the *Village of Woodridge* Code, in that said Defendant, the owner of *7660 Chicago Avenue, Woodridge, Illinois,* failed to provide the required clearance to combustible materials in that *there are cardboard boxes surrounding the furnace.*

Karyn Byrne

Complainant

Sworn to and Subscribed before Me
This *3rd* Day of *January, 2007*

_____
Notary Public

FAILURE TO MAINTAIN SAFETY CONTROLS—IPMC 603.4

STATE OF *ILLINOIS*

COUNTY OF DUPAGE

VILLAGE OF WOODRIDGE

v.

NAME: *PAT FOLEY*

ADDRESS: *7760 County Line Lane*

CITY: *Woodridge, Illinois 60517*

The undersigned says that on or about *December 26, 2006*, at or about *10:00 a.m.* the Defendant did unlawfully commit the offense of **Failure to Maintain Safety Controls** in violation of **2006 IPMC-603.4** as amended and adopted by reference in Section *8-1J-1(A)* of the *Village of Woodridge* Code, in that said Defendant, the owner of *4000 Quincy St., Woodridge, Illinois*, failed to maintain the safety controls on the *boiler* located therein in that *the safety valve has been bypassed.*

Karyn Byrne

Complainant

Sworn to and Subscribed before Me

This *3rd* Day of *January, 2007*

_____

Notary Public

- -

FAILURE TO PROVIDE COMBUSTION AIR—IPMC 603.5

STATE OF *ILLINOIS*
COUNTY OF DUPAGE
VILLAGE OF WOODRIDGE
v.
NAME: *PAT FOLEY*
ADDRESS: *7760 County Line Lane*
CITY: *Woodridge, Illinois 60517*

The undersigned says that on or about *December 26, 2006*, at or about *10:00 a.m.* the Defendant did unlawfully commit the offense of **Failure to Provide Combustion Air** in violation of *2006* **IPMC-603.5** as amended and adopted by reference in Section *8-1J-1(A)* of the *Village of Woodridge* Code, in that said Defendant, the owner of *4000 Quincy St., Woodridge, Illinois*, failed to provide a supply of air for complete combustion of the fuel, and for ventilations of the space containing the fuel-burning equipment in that *the furnace was in a closet with a solid door without any venting*.

Karyn Byrne

Complainant

Sworn to and Subscribed before Me
This *3rd* Day of *January, 2007*

_____

Notary Public

INSTALLING AN IMPROPER ENERGY CONSERVATION DEVICE—IPMC 603.6

STATE OF *ILLINOIS*
COUNTY OF DUPAGE
VILLAGE OF WOODRIDGE
v.
NAME: *PAT FOLEY*
ADDRESS: *7760 County Line Lane*
CITY: *Woodridge, Illinois 60517*

The undersigned says that on or about *December 26, 2006*, at or about *10:00 a.m.* the Defendant did unlawfully commit the offense of Installing an **Improper Energy Conservation Device** in violation of *2006* **IPMC-603.6** as amended and adopted by reference in Section *8-1J-1(A)* of the *Village of Woodridge* Code, in that said Defendant, the owner of *4000 Quincy St., Woodridge, Illinois*, installed a device intended to reduce fuel consumption on the *gas dryer,* a fuel-burning appliance,* located on the premises, which device was not labeled as being proper for such purpose and without specific approval in that *he installed a dryer box on a gas dryer.*

Karyn Byrne

Complainant

Sworn to and Subscribed before Me
This *3rd* Day of *January, 2007*

_____

Notary Public

*Or, to the fuel supply line thereto, or to the vent outlet or vent piping therefrom.

IPMC 604   Electrical Facilities Violations

FAILURE TO PROVIDE ELECTRICAL SYSTEM—IPMC 604.1

STATE OF *ILLINOIS*
COUNTY OF DUPAGE
VILLAGE OF WOODRIDGE
v.
NAME: *PAT FOLEY*
ADDRESS: *7760 County Line Lane*
CITY: *Woodridge, Illinois 60517*

The undersigned says that on or about *December 26, 2006*, at or about *10:00 a.m.* the Defendant did unlawfully commit the offense of **Failure to Provide Electrical System** in violation of *2006* **IPMC-604.1** as amended and adopted by reference in Section *8-1J-1(A)* of the *Village of Woodridge* Code, in that said Defendant, the owner of *4000 Quincy St., Woodridge, Illinois*, failed to provide an electrical system in compliance with the requirements of Section 605 as amended and adopted by reference in Section *8-1J-1(A)* of the *Village of Woodridge* Code in that *the fuse box is insufficient for the electrical load of the building.*

Karyn Byrne

Complainant

Sworn to and Subscribed before Me
This *3rd* Day of *January, 2007*

_____

Notary Public

FAILURE TO PROVIDE PROPER ELECTRICAL SERVICE—IPMC 604.2

STATE OF *ILLINOIS*
COUNTY OF DUPAGE
VILLAGE OF WOODRIDGE
v.
NAME: *PAT FOLEY*
ADDRESS: *7760 County Line Lane*
CITY: *Woodridge, Illinois 60517*

The undersigned says that on or about *December 26, 2006*, at or about *10:00 a.m.* the Defendant did unlawfully commit the offense of **Failure to Provide Proper Electrical Service** in violation of *2006* **IPMC-604.2** as amended and adopted by reference in Section *8-1J-1(A)* of the *Village of Woodridge* Code, in that said Defendant, the owner of *7660 Chicago Ave., Woodridge, Illinois*, failed to provide proper electrical service in that the service for the dwelling unit was less than the three-wire, 120/240 volt, single phase electrical service with a rating of not less than 60 amperes required in *that it was only 30 amperes.*

Karyn Byrne

Complainant

Sworn to and Subscribed before Me
This *3rd* Day of *February, 2007*

_____

Notary Public

- -
FAILURE TO CORRECT ELECTRICAL SYSTEM HAZARDS—IPMC 604.3

STATE OF *ILLINOIS*
COUNTY OF DUPAGE
VILLAGE OF WOODRIDGE
v.
NAME: *PAT FOLEY*
ADDRESS: *7760 County Line Lane*
CITY: *Woodridge, Illinois 60517*

The undersigned says that on or about *February 3, 2007*, at or about *10:00 a.m.* the Defendant did unlawfully commit the offense of **Failure to Correct Electrical System Hazards** in violation of *2006* **IPMC-604.3** as amended and adopted by reference in Section *8-1J-1(A)* of the *Village of Woodridge* Code, in that said Defendant, the owner of *7660 Chicago Ave., Woodridge, Illinois*, after being ordered to correct the electrical hazards on the premises by the code official on December 26, 2006, failed to correct the electrical hazards, being ***inadequate service,**** in that pennies being used in the fuse box to prevent the fuses from blowing, and improper use of extension cords.*

Karyn Byrne

Complainant

Sworn to and Subscribed before Me
This *3rd* Day of *February, 2007*

_____

Notary Public

*Or, improper fusing, insufficient receptacle and lighting outlets, improper wiring or installation, deterioration or damage, or similar reasons.

IPMC 605   Electrical Equipment Violations

IPMC 605.1—Installation

Sample Complaint—Failure to Properly Install Electrical Wiring   248

IPMC 605.2—Receptacles

Table 6-2—Required Electrical Receptacles   248

Sample Complaint—Failure to have Necessary Electrical Receptacle   249

IPMC 605.3—Luminaires

Sample Complaint—Failure to have Luminaire—2006 IPMC   250

Sample Complaint—Failure to have Lighting Fixture—2003 IPMC   251

- -

FAILURE TO PROPERLY INSTALL* ELECTRICAL WIRING**—IPMC 605.1

STATE OF *ILLINOIS*
COUNTY OF DUPAGE
VILLAGE OF WOODRIDGE
v.
NAME: *PAT FOLEY*
ADDRESS: *7760 County Line Lane*
CITY: *Woodridge, Illinois 60517*

The undersigned says that on or about *December 26, 2006*, at or about *10:00 a.m.* the Defendant did unlawfully commit the offense of **Failure to Properly Install* Electrical Wiring**** in violation of *2006* **IPMC-605.1** as amended and adopted by reference in Section *8-1J-1(A)* of the *Village of Woodridge* Code, in that said Defendant, the owner of *7660 Chicago Ave., Woodridge, Illinois*, improperly installed* electrical wiring** in an unsafe and unapproved manner on the premises *in the living room* in that *he used an extension cord as permanent wiring by plastering it into a wall and using it as an electrical receptacle.*

Karyn Byrne

Complainant

Sworn to and Subscribed before Me
This *3rd* Day of *February, 2007*

_____
Notary Public

  *Or, maintained.
**Or, equipment or appliances.

TABLE 6-2

- -

REQUIRED ELECTRICAL RECEPTACLES—IPMC 605.2

Location	Requirement
Habitable space in dwelling	At least 2 separate and remote receptacles
Laundry area	At least 1 grounded-type receptacle or receptacle with a ground fault circuit interrupter
Bathroom	At least 1 receptacle
	Note: New bathroom receptacle outlets must have ground fault circuit interrupter protection

SAMPLE COMPLAINT

FAILURE TO HAVE NECESSARY ELECTRICAL RECEPTACLE—IPMC 605.2

STATE OF *ILLINOIS*
COUNTY OF DUPAGE
VILLAGE OF WOODRIDGE
v.
NAME: *PAT FOLEY*
ADDRESS: *7760 County Line Lane*
CITY: *Woodridge, Illinois 60517*

The undersigned says that on or about *December 26, 2006*, at or about *10:00 a.m.* the Defendant did unlawfully commit the offense of **Failure to Have Necessary Electrical Receptacle** in violation of ***2006* IPMC-605.2** as amended and adopted by reference in Section *8-1J-1(A)* of the *Village of Woodridge* Code, in that said Defendant, the owner of *7660 Chicago Ave., Woodridge, Illinois,* had a bathroom receptacle outlet without ground fault circuit interrupter protection.*

Karyn Byrne

Complainant

Sworn to and Subscribed before Me
This *3rd* Day of *February, 2007*

_____

Notary Public

*Or, failed to have at least two separate and remote receptacles in a habitable space, being (*name of room*), or failed to have at least one grounded-type receptacle in the laundry room or failed to have at least one receptacle in the bathroom.

FAILURE TO HAVE LUMINAIRE—2006 IPMC 605.3

STATE OF *ILLINOIS*
COUNTY OF DUPAGE
VILLAGE OF WOODRIDGE
v.
NAME: *PAT FOLEY*
ADDRESS: *7760 County Line Lane*
CITY: *Woodridge, Illinois 60517*

The undersigned says that on or about *December 26, 2006*, at or about *10:00 a.m.* the Defendant did unlawfully commit the offense of **Failure to Have Luminaire** in violation of *2006* **IPMC-605.3** as amended and adopted by reference in Section *8-1J-1(A)* of the *Village of Woodridge* Code, in that said Defendant, the owner of *7660 Chicago Ave., Woodridge, Illinois*, did not provide an electric luminaire in the furnace room.*

Karyn Byrne
Complainant

Sworn to and Subscribed before Me
This *3rd* Day of *February, 2007*

_____
Notary Public

*Or, public hall, interior stairway, toilet room, kitchen, bathroom, laundry room, or boiler room.

COMMENT: The 2006 IPMC changes the language to "luminaires" from the former term "light fixtures."

SAMPLE COMPLAINT

FAILURE TO HAVE LIGHTING FIXTURE—2003 IPMC 605.3

STATE OF *ILLINOIS*
COUNTY OF DUPAGE
VILLAGE OF WOODRIDGE
v.
NAME: *PAT FOLEY*
ADDRESS: *7760 County Line Lane*
CITY: *Woodridge, Illinois 60517*

The undersigned says that on or about *December 26, 2006*, at or about *10:00 a.m.* the Defendant did unlawfully commit the offense of **Failure to Have Lighting Fixture** in violation of *2003* **IPMC-605.3** as amended and adopted by reference in Section *8-1J-1(A)* of the *Village of Woodridge* Code, in that said Defendant, the owner of *7660 Chicago Ave., Woodridge, Illinois*, did not provide an electric lighting fixture in the furnace room.*

Karyn Byrne

Complainant

Sworn to and Subscribed before Me
This *3rd* Day of *February, 2007*

_____

Notary Public

*Or, public hall, interior stairway, toilet room, kitchen, bathroom, laundry room, or boiler room.

IPMC 606   Elevator, Escalator, and Dumbwaiter Violations

SAMPLE COMPLAINT

FAILURE TO MAINTAIN ELEVATOR*—2006 IPMC 606.1

STATE OF *ILLINOIS*
COUNTY OF DUPAGE
VILLAGE OF WOODRIDGE
v.
NAME: *MC GINNIS DEPARMENT STORE LTD.*
ADDRESS: *200 Lincoln Blvd.*
CITY: *Woodridge, Illinois 60517*

The undersigned says that on or about *October 3, 2006*, at or about *2:00 p.m.* the Defendant did unlawfully commit the offense of **Failure to Maintain Elevator*** in violation of *2006* **IPMC-606.1** as amended and adopted by reference in Section *8-1J-1(A)* of the *Village of Woodridge* Code, in that said Defendant, the owner of *200 Lincoln Blvd., Woodridge, Illinois*, did fail to maintain the elevator* located on the premises in compliance with ASME A17.1-2000 in that *the elevator kept stopping between floors.*

Karyn Byrne

Complainant

Sworn to and Subscribed before Me
This *28th* Day of *October, 2006*

_____

Notary Public

*Or, escalator or dumbwaiter.

COMMENT: The 2006 IPMC adds ASME A17.1—2000 Safety Code for Elevators and Escalators as the specific standard for the maintenance, inspecting, and testing of elevators, dumbwaiters, and escalators.

FAILURE TO MAINTAIN ELEVATOR*—2003 IPMC 606.1

STATE OF *ILLINOIS*
COUNTY OF DUPAGE
VILLAGE OF WOODRIDGE
v.
NAME: *MC GINNIS DEPARMENT STORE LTD.*
ADDRESS: *200 Lincoln Blvd.*
CITY: *Woodridge, Illinois 60517*

The undersigned says that on or about *October 3, 2006,* at or about *2:00 p.m.* the Defendant did unlawfully commit the offense of **Failure to Maintain Elevator*** in violation of *2003* **IPMC-606.1** as amended and adopted by reference in Section *8-1J-1(A)* of the *Village of Woodridge* Code, in that said Defendant, the owner of *200 Lincoln Blvd., Woodridge, Illinois,* did fail to maintain the elevator* located on the premises so as to operate properly** in that *the elevator kept stopping between floors.*

Karyp Byrne

Complainant

Sworn to and Subscribed before Me
This *28th* Day of *October, 2006*

_____

Notary Public

 *Or, escalator or dumbwaiter.

**Or, so as to sustain safely all imposed loads or free from physical and fire hazards.

SAMPLE COMPLAINT

FAILURE TO DISPLAY CERTIFICATE OF ELEVATOR* INSPECTION—IPMC 606.1

STATE OF *ILLINOIS*
COUNTY OF DUPAGE
VILLAGE OF WOODRIDGE
v.
NAME: *MC GINNIS DEPARMENT STORE LTD.*
ADDRESS: *200 Lincoln Blvd.*
CITY: *Woodridge, Illinois 60517*

The undersigned says that on or about *October 3, 2006*, at or about *2:00 p.m.* the Defendant did unlawfully commit the offense of **Failure to Display Certificate of Elevator* Inspection** in violation of *2006* **IPMC-606.1** as amended and adopted by reference in Section *8-1J-1(A)* of the *Village of Woodridge* Code, in that said Defendant, the owner of *200 Lincoln Blvd., Woodridge, Illinois*, did fail to display the most current certificate of elevator inspection within the elevator* or make it available to the public for inspection in the office of the building operator.

Karyn Byrne

Complainant

Sworn to and Subscribed before Me
This *28th* Day of *October, 2006*

_____

Notary Public

*Or, attached to the escalator or dumbwaiter.

SAMPLE COMPLAINT

FAILURE TO INSPECT ELEVATOR*—2006 IPMC 606.1

STATE OF *ILLINOIS*
COUNTY OF DUPAGE
VILLAGE OF WOODRIDGE
v.
NAME: *MC GINNIS DEPARMENT STORE LTD.*
ADDRESS: *200 Lincoln Blvd.*
CITY: *Woodridge, Illinois 60517*

The undersigned says that on or about *October 3, 2006*, at or about *2:00 p.m.* the Defendant did unlawfully commit the offense of **Failure to Inspect Elevator*** in violation of ***2006* IPMC-606.1** as amended and adopted by reference in Section *8-1J-1(A)* of the *Village of Woodridge* Code, in that said Defendant, the owner of *200 Lincoln Blvd., Woodridge, Illinois*, did fail to perform the required inspection and tests upon the elevator* on the premises in compliance with the intervals listed in ASME A17.1, Appendix N, in that *3 months have elapsed since the required inspection and testing time.*

Karyn Byrne
Complainant

Sworn to and Subscribed before Me
This *28th* Day of *October, 2006*

_____
Notary Public

*Or, escalator or dumbwaiter.

COMMENT: The 2006 IPMC requires inspections and tests of elevators, escalators, and dumbwaiters at intervals set forth in the ASME A17.1-2000 unless the local jurisdiction specifies something different.

SAMPLE COMPLAINT

FAILURE TO MAINTAIN PASSENGER ELEVATOR—IPMC 606.2

STATE OF *ILLINOIS*
COUNTY OF DUPAGE
VILLAGE OF WOODRIDGE
v.
NAME: *MC GINNIS DEPARMENT STORE LTD.*
ADDRESS: *200 Lincoln Blvd.*
CITY: *Woodridge, Illinois 60517*

The undersigned says that on or about *October 3, 2006*, at or about *2:00 p.m.* the Defendant did unlawfully commit the offense of **Failure to Maintain Passenger Elevator** in violation of **2006 IPMC-606.2** as amended and adopted by reference in Section *8-1J-1(A)* of the *Village of Woodridge* Code, in that said Defendant, the owner of *200 Lincoln Blvd., Woodridge, Illinois,* did fail to maintain at least one elevator in operation at all times during the time the building was occupied by *employees and patrons.*

Karyn Byrne

Complainant

Sworn to and Subscribed before Me
This *28th* Day of *October, 2006*

_____

Notary Public

COMMENT: In buildings with only one elevator, the elevator may be temporarily out of service for testing or servicing.

IPMC 607   Duct System Violations

IPMC 607.1—General

Sample Complaint—Failure to Maintain
Duct System   259

SAMPLE COMPLAINT

FAILURE TO MAINTAIN DUCT SYSTEM—IPMC 607.1

STATE OF *ILLINOIS*
COUNTY OF DUPAGE
VILLAGE OF WOODRIDGE
v.
NAME: *PAT FOLEY*
ADDRESS: *7760 County Line Lane*
CITY: *Woodridge, Illinois 60517*

The undersigned says that on or about *December 26, 2006*, at or about *10:00 a.m.* the Defendant did unlawfully commit the offense of **Failure to Properly Maintain Duct System** in violation of *2006* **IPMC-607.1** as amended and adopted by reference in Section *8-1J-1(A)* of the *Village of Woodridge* Code, in that said Defendant, the owner of *7660 Chicago Ave., Woodridge, Illinois*, failed to maintain the duct system in the home free from obstructions and capable of performing the required function in that *the duct from the gas dryer to the outside was blocked by lint and a dead squirrel, thereby creating a fire hazard.*

Karyn Byrne

Complainant

Sworn to and Subscribed before Me
This *30th* Day of *December, 2006*

_____

Notary Public

Fire Safety Violations

CHAPTER 7 covers violations in existing structures which fail to meet the minimum standards for fire safety requirements. The chapter prescribes minimum standards for fire safety in structures and exterior premises. It also regulates fire safety facilities and equipment to be provided in those structures.

IPMC 701   General Violations

IPMC 701.2—Responsibility

SAMPLE COMPLAINT

FAILURE TO MAINTAIN* FIRE SAFETY EQUIPMENT**—IPMC 701.2

STATE OF *ILLINOIS*
COUNTY OF DUPAGE
VILLAGE OF WOODRIDGE
v.
NAME: *TIM HALIK*
ADDRESS: *6560 Hollywood Blvd.*
CITY: *Woodridge, Illinois 60517*

The undersigned says that on or about *June 22, 2006*, at or about *10:00 a.m.* the Defendant did unlawfully commit the offense of **Failure to Maintain* Fire Safety Equipment**** in violation of *2006* **IPMC-701.2** as amended and adopted by reference in Section *8-1J-1(A)* of the *Village of Woodridge* Code, in that said Defendant, the owner of *6560 Hollywood Blvd., Woodridge, Illinois*, failed to maintain* fire safety equipment** on the premises in violation of *2006* IPMC *704.1 Woodridge, Illinois* in that *the smoke detectors in the bedrooms have no batteries.*

Pat Kenny

Complainant

Sworn to and Subscribed before Me
This *27th* Day of *July, 2006*

_____

Notary Public

 *Or, provide.
**Or, facilities.

OCCUPYING PREMISES IN VIOLATION OF FIRE SAFETY REQUIREMENTS—IPMC 701.2

STATE OF *ILLINOIS*
COUNTY OF DUPAGE
VILLAGE OF WOODRIDGE
v.
NAME: *TIM HALIK*
ADDRESS: *6560 Hollywood Blvd.*
CITY: *Woodridge, Illinois 60517*

The undersigned says that on or about *June 22, 2006*, at or about *10:00 a.m.* the Defendant did unlawfully commit the offense of **Occupying Premises in Violation of Fire Safety Requirements** in violation of *2006* **IPMC-701.2** as amended and adopted by reference in Section *8-1J-1(A)* of the *Village of Woodridge* Code, in that said Defendant, the owner-occupant of *6560 Hollywood Blvd., Woodridge, Illinois,* occupied said premises at a time when the premises were in violation of 2006 IPMC *702.3* in that *a key was needed from the inside to open the back door of the house.*

Pat Kenny

Complainant

Sworn to and Subscribed before Me
This *27th* Day of *July, 2006*

_____
Notary Public

SAMPLE COMPLAINT

ALLOWING PERSONS TO OCCUPY PREMISES IN VIOLATION OF FIRE SAFETY REQUIREMENTS—IPMC 701.2

STATE OF *ILLINOIS*

COUNTY OF DUPAGE

VILLAGE OF WOODRIDGE

v.

NAME: *TIM HALIK*

ADDRESS: *6560 Hollywood Blvd.*

CITY: *Woodridge, Illinois 60517*

The undersigned says that on or about *June 22, 2006*, at or about *10:00 a.m.* the Defendant did unlawfully commit the offense of **Allowing Persons to Occupy Premises in Violation of Fire Safety Requirements** in violation of *2006* **IPMC-701.2** as amended and adopted by reference in Section *8-1J-1(A)* of the *Village of Woodridge* Code, in that said Defendant, the owner of *7660 Chicago Avenue, Woodridge, Illinois,* allowed persons to occupy said premises at a time when the premises were in violation of 2006 IPMC *702.4* in that *the defendant allowed persons to occupy the basement of the house at a time when the emergency escape windows were blocked by plywood.*

Pat Kenny

Complainant

Sworn to and Subscribed before Me

This *27th* Day of *July, 2006*

_____

Notary Public

IPMC 702   Means of Egress Violations

FAILURE TO PROVIDE PROPER MEANS OF EGRESS—IPMC 702.1

STATE OF *ILLINOIS*
COUNTY OF DUPAGE
VILLAGE OF WOODRIDGE
v.
NAME: *KYLE'S CHICKEN SHACK, INC.*
ADDRESS: *109 Fremont St.*
CITY: *Woodridge, Illinois 60517*

The undersigned says that on or about *July 18, 2006*, at or about *10:00 a.m.* the Defendant did unlawfully commit the offense of **Failure to Provide Proper Means of Egress** in violation of ***2006* IPMC-702.1** as amended and adopted by reference in Section *8-1J-1(A)* of the *Village of Woodridge* Code, in that said Defendant, the owner of *109 Fremont St., Woodridge, Illinois*, failed to provide a safe, continuous path of travel from the *kitchen* in the building to the public way in that *the corridor from the kitchen to the back door is crowded with boxes and supplies, thereby hindering access to the rear exit.*

Pat Kenny

Complainant

Sworn to and Subscribed before Me
This *15th* Day of *August, 2006*

_____
Notary Public

OBSTRUCTED AISLE WIDTH—IPMC 702.2

STATE OF *ILLINOIS*
COUNTY OF DUPAGE
VILLAGE OF WOODRIDGE
v.
NAME: *MC GINNIS DEVELOPMENT LLC*
ADDRESS: *200 Lincoln Blvd.*
CITY: *Woodridge, Illinois 60517*

The undersigned says that on or about *October 3, 2006,* at or about *8:00 p.m.* the Defendant did unlawfully commit the offense of **Obstructed Aisle Width** in violation of *2006* **IPMC-702.2** as amended and adopted by reference in Section *8-1J-1(A)* of the *Village of Woodridge* Code, in that said Defendant, the owner of *200 Lincoln Blvd., Woodridge, Illinois,* allowed obstructed aisles in the building, in that *chairs and music equipment occupied the required width of the aisles in the concert facility,* thereby reducing the required width for aisles under the International Fire Code.

Pat Kenny

Complainant

Sworn to and Subscribed before Me
This *28th* Day of *October, 2006*

_____

Notary Public

SAMPLE COMPLAINT

LOCKED MEANS OF EGRESS—IPMC 702.3

STATE OF *ILLINOIS*
COUNTY OF DUPAGE
VILLAGE OF WOODRIDGE
v.
NAME: *MC GINNIS DEVELOPMENT LLC*
ADDRESS: *200 Lincoln Blvd.*
CITY: *Woodridge, Illinois 60517*

The undersigned says that on or about *October 3, 2006*, at or about *8:00 p.m.* the Defendant did unlawfully commit the offense of **Locked Means of Egress** in violation of *2006* **IPMC-702.3** as amended and adopted by reference in Section *8-1J-1(A) Village of Woodridge* Code, in that said Defendant, the owner of *200 Lincoln Blvd., Woodridge, Illinois*, failed to have a means of egress door readily openable from the side from which egress is to be made without the need for keys, special knowledge, or effort in that *the rear door to the concert facility is chained shut.*

Pat Kenny
_____
Complainant

Sworn to and Subscribed before Me
This *5th* Day of *October, 2006*

_____
Notary Public

FAILURE TO MAINTAIN EMERGENCY ESCAPE OPENINGS—IPMC 702.4

STATE OF *ILLINOIS*
COUNTY OF DUPAGE
VILLAGE OF WOODRIDGE
v.
NAME: *TIM HALIK*
ADDRESS: *6560 Hollywood Blvd.*
CITY: *Woodridge, Illinois 60517*

The undersigned says that on or about *June 22, 2006*, at or about *10:00 a.m.* the Defendant did unlawfully commit the offense of **Failure to Maintain Emergency Escape Openings** in violation of *2006* **IPMC-702.4** as amended and adopted by reference in Section *8-1J-1(A)* of the *Village of Woodridge* Code, in that said Defendant, the owner of *6560 Hollywood Blvd., Woodridge, Illinois*, failed to maintain a required emergency escape opening, being *the escape window in the basement* required by the code in effect at the time of construction in that *the defendant allowed the emergency escape window to be blocked by plywood.*

Pat Kenny

Complainant

Sworn to and Subscribed before Me
This *27th* Day of *July, 2006*

_____

Notary Public

SAMPLE COMPLAINT

INOPERABLE EMERGENCY ESCAPE OPENINGS—IPMC 702.4

STATE OF *ILLINOIS*
COUNTY OF DUPAGE
VILLAGE OF WOODRIDGE
v.
NAME: *TIM HALIK*
ADDRESS: *6560 Hollywood Blvd.*
CITY: *Woodridge, Illinois 60517*

The undersigned says that on or about *June 22, 2006*, at or about *10:00 a.m.* the Defendant did unlawfully commit the offense of **Inoperable Emergency Escape Openings** in violation of ***2006* IPMC-702.4** as amended and adopted by reference in Section *8-1J-1(A)* of the *Village of Woodridge*, in that said Defendant, the owner of *6560 Hollywood Blvd., Woodridge, Illinois*, failed to have an operational emergency escape opening, being *the windows in the living room* in that *burglar bars over the windows required a key to open them from the inside.*

Pat Kenny

Complainant

Sworn to and Subscribed before Me
This *27th* Day of *July, 2006*

_____

Notary Public

COMMENT: Bars, grilles, grates or similar devices are permitted to be placed over emergency escape and rescue openings if the minimum net clear opening size complies with the code that was in effect at the time of construction and the devices are releasable or removable from the inside without the use of a key, tool or force greater than that which is required for normal operation of the escape and rescue opening.

IPMC 703   Fire-Resistance Ratings Violations

SAMPLE COMPLAINT

FAILURE TO MAINTAIN FIRE-RESISTANCE RATINGS—IPMC 703.1

STATE OF *ILLINOIS*
COUNTY OF DUPAGE
VILLAGE OF WOODRIDGE

v.

NAME: *MC GINNIS DEVELOPMENT LLC*
ADDRESS: *200 Lincoln Blvd.*
CITY: *Woodridge, Illinois 60517*

The undersigned says that on or about *October 3, 2006*, at or about *8:00 p.m.* the Defendant did unlawfully commit the offense of **Failure to Maintain Fire-Resistance Ratings** in violation of *2006* **IPMC-703.1** as amended and adopted by reference in Section *8-1J-1(A)* of the *Village of Woodridge*, in that said Defendant, the owner of *200 Lincoln Blvd., Woodridge, Illinois*, failed to maintain the fire-resistance rating of the *corridor* wall* in that *duct work penetrated the wall without proper sealing.*

Pat Kenny

Complainant

Sworn to and Subscribed before Me
This *5th* Day of *October, 2006*

_____

Notary Public

*Or, fire stop, shaft enclosure, partition, or floor.

SAMPLE COMPLAINT

FAILURE TO MAINTAIN REQUIRED OPENING PROTECTIVES—IPMC 703.2

STATE OF *ILLINOIS*
COUNTY OF DUPAGE
VILLAGE OF WOODRIDGE
v.
NAME: *MC GINNIS DEVELOPMENT LLC*
ADDRESS: *200 Lincoln Blvd.*
CITY: *Woodridge, Illinois 60517*

The undersigned says that on or about *October 3, 2006*, at or about *8:00 p.m.* the Defendant did unlawfully commit the offense of **Failure to Maintain Required Opening Protectives** in violation of *2006* **IPMC-703.2** as amended and adopted by reference in Section *8-1J-1(A)* of the *Village of Woodridge* Code, in that said Defendant, the owner of *200 Lincoln Blvd., Woodridge, Illinois*, failed to maintain the required opening protective, being the *fire door* in that *the entire door assembly does not have a one hour fire rating.*

Pat Kenny
Complainant

Sworn to and Subscribed before Me
This *5th* Day of *October, 2006*

_____
Notary Public

SAMPLE COMPLAINT

INOPERABLE* FIRE DOOR**—IPMC 703.2

STATE OF *ILLINOIS*
COUNTY OF *DUPAGE*
VILLAGE OF *WOODRIDGE*
v.
NAME: *MC GINNIS DEVELOPMENT LLC*
ADDRESS: *200 Lincoln Blvd.*
CITY: *Woodridge, Illinois 60517*

The undersigned says that on or about *October 3, 2006*, at or about *8:00 p.m.* the Defendant did unlawfully commit the offense of **Inoperable* Fire Door**** in violation of *2006* **IPMC-703.2** as amended and adopted by reference in Section *8-1J-1(A)* of the *Village of Woodridge* Code, in that said Defendant, the owner of *200 Lincoln Blvd., Woodridge, Illinois,* failed to maintain a fire door** in an operable condition*** in that *the door closer has been removed.*

Pat Kenny

Complainant

Sworn to and Subscribed before Me
This *5th* Day of *October, 2006*

_____

Notary Public

   *Or, Fire Door in an Obstructed Condition.
  **Or, smoke barrier door.
***Or, in an unobstructed condition.

IPMC 704   Fire Protection System Violations

SAMPLE COMPLAINT

FAILURE TO MAINTAIN FIRE PROTECTION SYSTEM—IPMC 704.1

STATE OF *ILLINOIS*
COUNTY OF DUPAGE
VILLAGE OF WOODRIDGE
v.
NAME: *MC GINNIS DEVELOPMENT LLC*
ADDRESS: *200 Lincoln Blvd.*
CITY: *Woodridge, Illinois 60517*

The undersigned says that on or about *October 3, 2006*, at or about *8:00 p.m.* the Defendant did unlawfully commit the offense of **Failure to Maintain Fire Protection System** in violation of *2006* **IPMC-704.1** as amended and adopted by reference in Section *8-1J-1(A)* of the *Village of Woodridge* Code, in that said Defendant, the owner of *200 Lincoln Blvd., Woodridge, Illinois,* failed to maintain the fire protection system, being *the fire alarm system,** in an operable condition* in that *the fire alarm system is not connected to the control panel.*

Pat Kenny

Complainant

Sworn to and Subscribed before Me
This *5th* Day of *October, 2006*

_____

Notary Public

*Or, device or equipment.

TABLE 7-1

REQUIRED SMOKE ALARMS

Location	Required?	Section
On the ceiling or wall outside of each separate sleeping area in the immediate vicinity of the bedrooms	Yes	704.2(1)
Sleeping rooms	Yes	704.2(2)
In each story within a dwelling unit including basements and cellars	Yes	704.2(3)
Crawl spaces and uninhabited attics	No	704.2(3)
Dwelling unit with split level and without intervening door	Yes	704.2(3)

Note: Smoke alarm installed on the upper level is sufficient for the adjacent lower level provided that the lower level is less than one full story below the upper level.

FAILURE TO INSTALL* SMOKE ALARM—IPMC 704.2

STATE OF *ILLINOIS*
COUNTY OF DUPAGE
VILLAGE OF WOODRIDGE
v.
NAME: *TIM HALIK*
ADDRESS: *6560 Hollywood Blvd.*
CITY: *Woodridge, Illinois 60517*

The undersigned says that on or about *June 22, 2006*, at or about *10:00 a.m.* the Defendant did unlawfully commit the offense of **Failure to Install* Smoke Alarm** in violation of ***2006 IPMC-704.12*** as amended and adopted by reference in Section *8-1J-1(A)* of the *Village of Woodridge* Code, in that said Defendant, the owner of *6560 Hollywood Blvd., Woodridge, Illinois*, failed to install* a single-station** smoke alarm in a Group R-2*** residence in that *there were no smoke detectors in the sleeping rooms.*

Pat Kenny

Complainant

Sworn to and Subscribed before Me
This *5th* Day of *October, 2006*

_____

Notary Public

   *Or, maintain.
  **Or, multiple-station alarm.
***Or, Group R-3, R-4 or a dwelling not regulated in a Group R occupancy.

Public Nuisance Violations

CHAPTER 8 covers public nuisance violations. Most of the local jurisdictions that have not adopted the International Property Maintenance Code use public nuisance ordinances to regulate the same types of violations found in the model code. The nature of public nuisance violations differs throughout the country given the varying conditions which exist in each locality, but the procedure for enforcing those violations is generally very similar. Consequently, this chapter can easily be adapted to local use.

Public Nuisance Violations

IN GENERAL

Black's Law Dictionary defines a nuisance as "Anything which is injurious to health or is indecent or offensive to the senses or an obstruction to the free use of property, so as to interfere with the comfortable enjoyment of life or property or which unlawfully obstructs the free passage or use, in the customary manner of any navigable lake or river, bay stream, canal or basin, or any public park, square, street to highway, is a nuisance . . . A public nuisance is one which affects an indefinite number of persons, or all the residents of a particular locality or all people coming within the extent of its range or operation, although the extent of the annoyance or damage inflicted upon individuals may be unequal."

Public Nuisance Violations

SAMPLE NUISANCE ORDINANCE

Most public nuisance ordinances contain a general definition of nuisance such as this one taken from the Code of Ordinances of the Village of Woodridge, Illinois:

1-3-2: NUISANCE: Anything offensive to the sensibilities of reasonable persons, or any act or activity creating a hazard which threatens the health and welfare of inhabitants of the village; or any activity which by its perpetuation can reasonably be said to have a detrimental effect on the property of a person or persons within the community.

The ordinances usually recite specific examples of what constitutes a nuisance such as in this provision, also taken from the ordinances of the Village of Woodridge:

4-1-1: NUISANCE: DEFINED AND PROHIBITED:

The following acts, conduct and conditions are hereby declared and defined to be nuisances, and when committed, performed and/or permitted to exist by any person within the territorial limits of the Village, are hereby declared to be unlawful and prohibited:

A. Common Law Nuisances: Any act or offense which is a nuisance according to the common law of the State of Illinois, or declared or defined to be a nuisance by the ordinances of the Village or of the statutes of the State of Illinois. In addition, the officials of the Municipality shall be authorized to abate any nuisance which, while not specifically defined within this Chapter, shall constitute the unreasonable, unwarrantable or unlawful use by a person or property, real or personal, or from his own improper, indecent or unlawful personal conduct which works in obstruction or injury to a right of another or of the public and produces such material annoyance, inconvenience, discomfort or hurt that the law will presume an actionable nuisance. Nuisances may be abated which are public or which are both public and private in nature.

B. General Conditions:

   1. Any thing, act, occupation or use of property, premises, equipment or structure adversely affecting the health, welfare and safety of individuals or of the public.

   2. Any act or omission, occupation and use of property, equipment or structure which shall in any way threaten the life or the use of property of individuals or of the public.

C. Dangerous, Unhealthful Buildings:

   1. All buildings, walls and other structures which have been damaged or neglected by fire, decay or otherwise so as to constitute a menace to the health, welfare or safety of individuals or of the public.

   2. To own, maintain or keep a dwelling unit unfit for human habitation or dangerous or detrimental to life, safety or health because of lack of repair, defects in the plumbing system, lighting or ventilation, the existence of contagious diseases or unsanitary conditions likely to cause sickness among persons residing in said premises, or residing in proximity thereof.

D. Animal Carcasses; Offal:

   1. To cause or suffer the carcass of any animal, or any offal, filth or noisome substance to be collected, deposited or to remain in any place under this ownership or control to the injury of others.

(Continued)

SAMPLE NUISANCE ORDINANCE (*CONTINUED*)

2. To throw or deposit any offal or any other offensive matter or the carcass of any dead animal in any watercourse, lake, pond, spring, well or common sewer, street or public highway.

E. Water Pollution: To corrupt or render unwholesome or impure the water of any well, spring, river, stream, pond or lake to the injury or prejudice of others.

F. Obstructing Public Ways: To obstruct or encroach upon public highways, private ways, streets, alleys, commons and landing places and ways to burying places.

G. Offensive Businesses: To erect, to continue or use any building or other place for the exercise of any trade, employment or manufacture, which, by occasioning noxious exhalations, offensive smells or otherwise is offensive or dangerous to the health of individuals or of the public.

H. Advertisements on Public or Private Property: To advertise wares or occupations by painting notices of the same on or affixing them to fences, walls, windows, building exteriors, utility poles, or on hydrants, other public or private property, or on rocks or other natural objects, without the consent of the owner, or if in the highway right of way or other public place, the permission of the Village authorities.

I. Harassment Regarding Real Estate Transactions: To harass, intimidate or threaten any person who is about to sell or lease or has sold or leased a residence or other real property, or is about to buy or lease or has bought or leased a residence or other real property, when the harassment, intimidation or threat relates to a person's attempt to sell, buy or lease a residence or other real property or refers to a person's sale, purchase or lease of a residence or other real property.

J. Unlawful Garbage and Refuse Deposits:

1. Public Property: To dump, abandon, deposit, dismantle or burn upon any public property or right of way, highway, park, street or parkway anywhere in the Village any trash, garbage, ashes, junk, junked or wrecked automobiles or parts thereof or miscellaneous waste.

2. Private Property: To dump, deposit or place any garbage, rubbish, trash or refuse upon real property owned by another without the consent of the owner or person in possession of such real property.

K. Open Junk Storage: To store, keep or maintain outside of a closed building, any junk, parts, machinery or equipment not in an operable condition, except as may be otherwise permitted by this code.

L. Inoperable Motor Vehicles: All inoperable motor vehicles, whether on public or private property. However, nothing in this subsection shall apply to any motor vehicle that is kept within a building when not in use, to historic vehicles over twenty five (25) years of age, or to a motor vehicle on the premises of a place of business engaged in the wrecking or junking of motor vehicles.

As used in this subsection, "inoperable motor vehicle" means any motor vehicle from which, for a period of at least seven (7) days, the engine, wheels or other parts have been removed, or on which the engine, wheels or other parts have been altered, damaged or otherwise so treated that the vehicle is incapable of being driven under its own motor power. "Inoperable motor vehicle" shall not include a motor vehicle which has been rendered temporarily incapable of being driven under its own motor power in order to perform ordinary services or repair operations.

M. Insect Infestations: All infestations of flies, fleas, roaches, lice, ticks, rats, mice, fly maggots, and mosquitoes and mosquito larvae. (1976 Code §10-100)

(Continued)

N. Noise:

1. To produce or permit to be produced, whether on private or public property, any offensive noise to the disturbance of the peace or quiet of any person residing in the vicinity.

2. For the owner and/or occupant of any premises to permit any of the following activities to be conducted on said premises from nine o'clock (9:00 p.m.) to seven o'clock (7:00 a.m.), within three hundred feet (300') of any residential district boundary:

   a. The collection, pick up or disposal of refuse, recyclable materials, empty containers, drums, etc., or any other reusable product or device, except when conducted entirely within a completely enclosed building

   b. The sweeping or cleaning of any parking lot or sidewalk by any mechanical means

   c. The shipping or receiving of any goods, merchandise or any other materials, except when conducted entirely within a completely enclosed building

   d. Any other outdoor maintenance activity requiring the use of any mechanical device

O. Dog Nuisances: Any dog which is defined as:

1. A vicious dog under Section 5-6A-1 of this code and which is not kept in compliance with Subsection 5-6A-5C of this code; or

2. A dangerous dog under Section 5-6A-1 of this code, which is not kept in compliance with Subsection 5-6A-5B of this code.

P. Tree and Plant Nuisances:

1. Disease Conditions: All trees, shrubs, vines, cuttings, scions, graphs, plants and plant parts and plant products in places within the village, infested with injurious insect pests or infected with plant diseases which are liable to spread to other plants, plant products or places to the injury thereof, or to the injury of man and animals, and all species and varieties of trees, shrubs, vines and other plants not essential to the welfare of the people of the village which may serve as a favorable host plant and promote the prevalence and abundance of insect pests and plant diseases, or any stage thereof, injurious to other plants essential to the welfare of the people of this village.

2. Dangerous Conditions; Encroachments: Any tree, shrub, or other planting:

   a. Which by its location or condition constitutes a threat to the safety or property of individuals or of the public

   b. Which obstructs or encroaches upon any street right of way, sidewalk, public property or any public or village utility lines or facilities

Q. Illegal Liquor Manufacture Or Sale: Every lot, parcel or tract of land, and every building, structure, tent, boat, wagon, vehicle, establishment or place whatsoever, together with all furniture, fixtures, ornaments and machinery located thereon, wherein there shall be conducted any unlawful manufacture, distribution, or sale of alcoholic liquor, or whereon or wherein there shall be kept, stored, concealed or allowed any alcoholic liquor intended for illegal sale or to be sold, disposed of or in any other manner used in violation of any of the provisions of 235 Illinois Compiled Statutes, as amended.

Most nuisance ordinances contain an abatement procedure which includes a notification procedure along with remedies if the problem is not corrected. An example of that type of ordinance, taken from the ordinances of the Village of Woodridge is set forth as follows:

(Continued)

SAMPLE NUISANCE ORDINANCE (*CONTINUED*)

4-1-2: ABATEMENT PROCEDURE:

A. Notice To Abate: Except where otherwise provided by this code, the village administrator, police chief, director of public services or director of building and zoning may serve or cause to be served a notice, in writing, upon the owner, agent, occupant or person in possession, charge or control of any lot, building or premises or item of personalty in or upon which any nuisance exists, requiring them, or either or all of them, to abate the same within a specified reasonable time, in such manner as the notice shall direct.

B. Abatement By Village: If the person so served and notified does not abate the nuisance within the specified reasonable time, the corporate authorities may proceed to abate the nuisance in any and all manners allowable by law, including, without limiting the generality thereof, the following:

   1. Seeking to impose a monetary penalty as defined by Subsection 4-1-3A of this chapter by instituting an ordinance enforcement action.

   2. Seeking to enjoin the continuation of the nuisance by the filing of a lawsuit in a court of competent jurisdiction.

C. Summary Abatement: Whenever, in the opinion of one of the village officers designated in subsection A of this section, the maintenance or continuation of a nuisance creates an eminent threat of serious injury to persons or serious damage to personal or real property, or if the nuisance can be abated summarily without or with only minor damage to the items or premises which are creating the nuisance, and the continuation of the nuisance poses a substantial threat of injury to persons or property or a substantial interference with the quiet enjoyment of life normally present in the community, such officer shall proceed to abate such nuisance; provided, further, that whenever the owner, occupant, agent or person in possession, in charge or control of the real or personal property which has become a nuisance is unknown or cannot readily be found, such officer may proceed to abate such nuisance without notice. Where the abatement of a nuisance requires continuing acts by the corporate authorities beyond the initial summary abatement and any other additional emergency abatements, it shall seek abatement of such nuisance on a permanent basis through judicial process as soon as reasonably practicable.

This chapter will provide examples of forms which are useful in enforcing these types of codes, such as a notice of abatement, and sample complaints for the most common type of violations involving structures and premises. To use these forms, the inspector should substitute the section numbers of the local jurisdiction for the ones used in the examples.

NOTICE OF ABATEMENT OF PUBLIC NUISANCE

April 2, 2006
Mr. John Fincham
6560 Hollywood Blvd.
Woodridge, Illinois 60517

> Re: *6560 Hollywood Blvd., Woodridge, Illinois*

Dear *Mr. Fincham*:

An inspection of your property at *6560 Hollywood Blvd., Woodridge, Illinois*, on *April 2, 2006* shows the following violations of the code of ordinances of the *Village of Woodridge*:

The house has been damaged by a fire and the roof is in danger of collapse so as to constitute a menace to the health, welfare or safety of individuals or of the public in violation of Section *4-1-1(C)(1)* of the Code of Ordinances of the *Village of Woodridge*.

The garage is being used for business purposes, in that meat is being deep fried in 5 gallon drums and then sold to the public, thereby being a threat to the life or use of the property of individuals or of the public in violation of Section *4-1-1(C)(2)* of the Code of Ordinances of the *Village of Woodridge*.

An abatement of these problems must be made by the close of business on *May 12, 2006* or a complaint will be filed against you in a court of local jurisdiction seeking a monetary penalty and/or an injunction to abate the continuation of the nuisance. All required permits must be obtained before any work commences.

If you fail to correct these violations, any action taken by the *Village of Woodridge*, the authority having jurisdiction, may be charged against the real estate upon which the structure is located and shall be a lien upon such real estate.

Please feel free to contact me to discuss this matter further.

Very truly yours,

Karyn Byrne

Deputy Code Official

MAINTAINING A PUBLIC NUISANCE (DISCHARGE OF SUMP PUMP WATER)

STATE OF *ILLINOIS*
COUNTY OF DUPAGE
VILLAGE OF WOODRIDGE
v.
NAME: *JOHN FINCHAM*
ADDRESS: *6560 Hollywood Blvd.*
CITY: *Woodridge, Illinois 60517*

The undersigned says that on or about *November 18, 2006* at or about *3:00 p.m.* the Defendant did unlawfully commit the offense of **Maintaining a Public Nuisance, Discharge of Sump Pump Water,** in violation of Section *4-1-1(A)* of the ordinances of the *Village of Woodridge*, in that said Defendant, the owner of *6560 Hollywood Blvd., Woodridge, Illinois, discharged his sump pump water onto the property of his next door neighbor and onto the public sidewalk and street, thereby creating a hazard when the neighbor's property floods and the water freezes on the public sidewalk and street.*

Karyn Byrne

Complainant

Sworn to and Subscribed before Me
This *30th* Day of *November, 2006*

_____

Notary Public

MAINTAINING A PUBLIC NUISANCE (STAGNANT WATER)

STATE OF *ILLINOIS*
COUNTY OF DUPAGE
VILLAGE OF WOODRIDGE
v.
NAME: *ROBERT MEYER*
ADDRESS: *4617 Carousel St.*
CITY: *Woodridge, Illinois 60517*

The undersigned says that on or about *August 15, 2006*, at or about *3:00 p.m.* the Defendant did unlawfully commit the offense of **Maintaining a Public Nuisance, Stagnant Water,** in violation of Section *4-1-1(B)(2)* of the ordinances of the *Village of Woodridge*, in that said Defendant, the owner of *4617 Carousel St., Woodridge, Illinois, allowed stagnant water to collect on the cover on his outdoor pool, thereby serving as a breeding ground for mosquitoes.*

Karyn Byrne

Complainant

Sworn to and Subscribed before Me
This *30th* Day of *August, 2006*

_____

Notary Public

MAINTAINING A PUBLIC NUISANCE (VACANT STRUCTURE)

STATE OF *ILLINOIS*
COUNTY OF DUPAGE
VILLAGE OF WOODRIDGE
v.
NAME: *ROBERT MEYER*
ADDRESS: *4617 Carousel St.*
CITY: *Woodridge, Illinois 60517*

The undersigned says that on or about *July 18, 2006*, at or about *3:00 p.m.* the Defendant did unlawfully commit the offense of **Maintaining a Public Nuisance being a Vacant Structure** in violation of Section *4-1-1(C)(1)* of the ordinances of the *Village of Woodridge*, in that said Defendant, the owner of *4617 Carousel St., Woodridge, Illinois*, maintained a public nuisance, the vacant structure at *4617 Carousel St., Woodridge, Illinois*, in that *the residence has broken windows, the front door is not secure, and the basement windows have been removed* so as to adversely affect the health, welfare, and safety of individuals or of the public.

Joan Rogers

Complainant

Sworn to and Subscribed before Me
This *30th* Day of *August, 2006*

_____

Notary Public

MAINTAINING A PUBLIC NUISANCE (BUILDING UNFIT FOR HUMAN HABITATION)

STATE OF *ILLINOIS*
COUNTY OF DUPAGE
VILLAGE OF WOODRIDGE
v.
NAME: *ROBERT MEYER*
ADDRESS: *4617 Carousel St.*
CITY: *Woodridge, Illinois 60517*

The undersigned says that on or about *April 16, 2006*, at or about *3:00 p.m.* the Defendant, the owner of *4617 Carousel St., Woodridge, Illinois*, did unlawfully commit the offense of **Maintaining a Public Nuisance, a Building Unfit for Human Habitation,** in violation of Section *4-1-1(C)(2)* of the ordinances of the *Village of Woodridge*, in that *the residence is filled with dog feces, has bare electrical wires in the room, boarded up windows, and no siding on the east side of the building*, thereby being unfit for human habitation, dangerous and detrimental to life, safety, or health because of lack of repair, and with unsanitary conditions likely to cause sickness among persons residing in said premises.

Joan Rogers
Complainant

Sworn to and Subscribed before Me
This *30th* Day of *April, 2006*

_____

Notary Public

MAINTAINING A PUBLIC NUISANCE (OBSTRUCTING A PUBLIC WAY)

STATE OF *ILLINOIS*
COUNTY OF DUPAGE
VILLAGE OF WOODRIDGE
v.
NAME: *TREVOR BISHOP*
ADDRESS: *129 Meadow St.*
CITY: *Woodridge, Illinois 60517*

The undersigned says that on or about *May 24, 2006*, at or about *2:00 p.m.* the Defendant, the owner of *129 Meadow St., Woodridge, Illinois,* did unlawfully commit the offense of **Maintaining a Public Nuisance, Obstructing a Public Way,** in violation of Section *4-4-1(F)* of the ordinances of the *Village of Woodridge* by obstructing a public way in that the Defendant permitted *the sidewalk on the premises to be ripped out and allowed construction material to be deposited on the public way, thereby blocking access by pedestrians and creating a public hazard.*

Don Lay
Complainant

Sworn to and Subscribed before Me
This *17th* Day of *June, 2006*

_____
Notary Public

MAINTAINING A PUBLIC NUISANCE (DUMPING GARBAGE OR REFUSE)

STATE OF *ILLINOIS*
COUNTY OF DUPAGE
VILLAGE OF WOODRIDGE
v.
NAME: *ROBERT MEYER*
ADDRESS: *4617 Carousel St.*
CITY: *Woodridge, Illinois 60517*

The undersigned says that on or about *April 2, 2006*, at or about *3:00 p.m.* the Defendant did unlawfully commit the offense of **Maintaining a Public Nuisance, Dumping Garbage or Refuse,** in violation of Section *4-1-1(J)(1)* of the ordinances of the *Village of Woodridge*, in that said Defendant deposited *numerous bags of trash and garbage* on public property, being *the Woodridge Festival Park.*

Joan Rogers

Complainant

Sworn to and Subscribed before Me
This *30th* Day of *April, 2006*

_____

Notary Public

MAINTAINING A PUBLIC NUISANCE (OPEN STORAGE OF JUNK)

STATE OF *ILLINOIS*
COUNTY OF DUPAGE
VILLAGE OF WOODRIDGE
v.
NAME: *ROBERT MEYER*
ADDRESS: *4617 Carousel St.*
CITY: *Woodridge, Illinois 60517*

The undersigned says that on or about *April 2, 2006*, at or about *3:00 p.m.* the Defendant did unlawfully commit the offense of **Maintaining a Public Nuisance, Open Storage of Junk,** in violation of Section *4-1-1(K)* of the ordinances of the *Village of Woodridge*, in that said Defendant, the owner of *4617 Carousel St., Woodridge, Illinois*, openly stored junk being *automobile parts, rusted machine parts, and rolls of rusted fencing in the backyard.*

Joan Rogers

Complainant

Sworn to and Subscribed before Me
This *30th* Day of *April, 2006*

_____

Notary Public

MAINTAINING A NUISANCE (INOPERABLE MOTOR VEHICLE)

STATE OF *ILLINOIS*
COUNTY OF DUPAGE
VILLAGE OF WOODRIDGE
v.
NAME: *ROBERT MEYER*
ADDRESS: *4617 Carousel St.*
CITY: *Woodridge, Illinois 60517*

The undersigned says that on or about *April 2, 2006*, at or about *3:00 p.m.* the Defendant did unlawfully commit the offense of **Maintaining a Public Nuisance, Inoperable Motor Vehicle,** in violation of Section *4-1-1(L)* of the ordinances of the *Village of Woodridge*; in that said Defendant, the owner of *4617 Carousel St., Woodridge, Illinois*, did maintain an inoperative motor vehicle on the premises, *a blue sedan in the driveway with no engine and two flat tires.*

Joan Rogers

Complainant

Sworn to and Subscribed before Me
This *30th* Day of *April, 2006*

_____

Notary Public

MAINTAINING A PUBLIC NUISANCE (INFESTED TREE)

STATE OF *ILLINOIS*
COUNTY OF DUPAGE
VILLAGE OF WOODRIDGE
v.
NAME: *TREVOR BISHOP*
ADDRESS: *129 Meadow St.*
CITY: *Woodridge, Illinois 60517*

The undersigned says that on or about *May 24, 2006,* at or about *2:00 p.m.* the Defendant, the owner of *129 Meadow St., Woodridge, Illinois,* did unlawfully commit the offense of **Maintaining a Public Nuisance, Infested Tree,** in violation of Section *4-4-1(P)(1)* of the ordinances of the *Village of Woodridge* in that *the elm tree on the property in the backyard is infected with Dutch Elm disease,* such disease being liable to spread to other trees.

Don Lay

Complainant

Sworn to and Subscribed before Me
This *17th* Day of *June, 2006*

_____

Notary Public

Index